Conditionally Toxic Proteins

Human health depends upon access to high-quality proteins for our diet and pharmaceutical use. This book deals with the nature of toxicity as it applies to proteins in food and drugs. Many proteins, such as glutens and allergens, are valuable food sources but toxic for sensitized individuals. Even proteins produced in the human body can become toxic through mutation or aggregation. For example, Alzheimer and Huntington diseases are characterized by plaques in the brain of insoluble protein aggregates. Paradoxically, even toxins produced by the most pathogenic microorganisms, such as Botox, have found use in the clinic and industry. This book discusses how many proteins, including interferons and cytokines, can be valuable therapeutics while still associated with autoimmune diseases, such as psoriasis or lupus erythematosus.

Key Features

- Shows why gluten proteins are different from all others and how this can account for symptoms throughout the body in those with celiac disease
- Illustrates that "multiple allergy syndrome" is real and related to similar allergens in extremely different plants
- Discusses the difficulties in using human proteins and cytokines as therapeutics
- Presents examples of how protein aggregates are vital to many physiological processes but also characterize renal and neurological syndromes
- Shows examples of protein toxins that have medical and industrial uses

Conditionally Toxic Proteins

Catherine H. Schein, MSc, PhD

CRC Press is an imprint of the
Taylor & Francis Group, an **informa** business

First edition published 2024
by CRC Press
6000 Broken Sound Parkway NW, Suite 300, Boca Raton, FL 33487-2742

and by CRC Press
4 Park Square, Milton Park, Abingdon, Oxon, OX14 4RN

CRC Press is an imprint of Taylor & Francis Group, LLC

© 2024 Catherine H. Schein

Reasonable efforts have been made to publish reliable data and information, but the author and publisher cannot assume responsibility for the validity of all materials or the consequences of their use. The authors and publishers have attempted to trace the copyright holders of all material reproduced in this publication and apologize to copyright holders if permission to publish in this form has not been obtained. If any copyright material has not been acknowledged please write and let us know so we may rectify in any future reprint.

Except as permitted under U.S. Copyright Law, no part of this book may be reprinted, reproduced, transmitted, or utilized in any form by any electronic, mechanical, or other means, now known or hereafter invented, including photocopying, microfilming, and recording, or in any information storage or retrieval system, without written permission from the publishers.

For permission to photocopy or use material electronically from this work, access www.copyright.com or contact the Copyright Clearance Center, Inc. (CCC), 222 Rosewood Drive, Danvers, MA 01923, 978-750-8400. For works that are not available on CCC please contact mpkbookspermissions@tandf.co.uk

Trademark notice: Product or corporate names may be trademarks or registered trademarks and are used only for identification and explanation without intent to infringe.

ISBN: 9781032366937 (hbk)
ISBN: 9781032366944 (pbk)
ISBN: 9781003333319 (ebk)

DOI: 10.1201/9781003333319

Typeset in Times
by KnowledgeWorks Global Ltd.

Contents

About the Author

Dr. Catherine H. Schein earned a BA from the University of Pennsylvania in Biochemistry, an MSc from the Massachusetts Institute of Technology in Biochemical Engineering and a PhD from the Swiss Federal Institute of Technology (Zürich) in Technical Microbiology, with thesis projects on hemoglobin and the extracellular proteins of pathogenic *Streptococci.* Her interest in conditionally toxic proteins began while in a blood lab at a major hospital, which introduced her to blood groupings, leukemias and other blood diseases. Watching a young cousin go into anaphylactic shock after eating a nut (he survived but does his own cooking) led to her work on determining why seemingly innocuous proteins in food and pollens can cause extremely negative reactions in certain people. She produced pharmaceutical proteins, interferons (IFN) and other cytokines, at the industrial scale, for treating infections and cancers. During this work, she developed methods to prevent the aggregation of proteins, which have been used to produce thousands of other proteins for research and medical use. After transitioning to academic research, she studied interactions of interferon with ribonucleases and other cellular proteins, inhibitors of viral proteins and bacterial virulence factors, and distinguishing properties of allergenic proteins. She also studied inhibitors of aggregation of human proteins that form plaques in the brain characteristic of Alzheimer and other neurodegenerative diseases.

This book reflects a central principle common to all these projects: that valuable pharmaceutical and food proteins can also be toxins, depending on context.

Acknowledgments

There are many people I need to thank for the experiences that helped me write this book. First, all my mentors and coworkers in my various positions, starting in the blood lab at Norfolk General Hospital to Switzerland to Texas to Florida and back, who taught me all the positive and negative aspects of the proteins we eat and use therapeutically. The chapters in this book all represent projects I have worked on throughout my scientific career, beginning with producing interferons and cytokines to designing vaccines and inhibitors of bacterial toxins and viruses.

I especially want to thank the readers of all the chapters of this book. My lab manger, Dr. Wendy S. Baker, took on the task of proof reading and has supported me throughout this effort. My good friend and one of the best violinists in the world, Sandra Goldberg, helped make the first chapters readable to those who want to better understand how the foods we eat and the air we breathe can contain conditional toxins. My friends in the science world, Dr. Steven L. Cheney, from the UTMB blood bank, and Dr. Balaji Krishnan, from the Mitchell Center for Neurodegenerative Disease, UTMB, helped with additions to Chapters 2 and 7. Dr. Penelope Kulla-Ness helped with her knowledge of industrial microbiology and fact checking of the dual uses of bacterial toxins.

Finally, I need to thank the editorial staff at Taylor & Francis, especially Riya Bhattacharya who helped correct many of the typos and word usages in the original manuscript.

Introduction
The Yin and Yang of Conditionally Toxic Proteins

> If you know your enemies and know yourself, you will not be put at risk even in a hundred battles.
>
> **Sun Tzu,** ***The Art of War***

Conditionally Toxic Proteins describes reasons why the proteins we eat, use cosmetically or for pharmaceutical purposes can be both functional and toxic. Many valuable and useful proteins can be our enemies, or friends, depending on many different factors. A protein's toxicity can be intrinsic, lie in alterations to its structure or result from its recognition as foreign by the immune system. Chapters in this book outline the medical reasons for using protein therapeutics while acknowledging and dealing with their associated toxicities. In some cases, as in supplying insulin to those with diabetes, or using accurately typed blood for transfusions, we win. In others, such as attempting to incorporate Interleukin 2 (IL-2) or tumor necrosis factor (TNF) into cancer therapies, we lose but go down trying. Many other cases require balancing associated toxicities with medically useful treatments.

Paradoxically, calling a protein *a toxin* seems to be an invitation to find a use for it that capitalizes precisely on its mechanism of action. Many toxic proteins have a clinical use, as examples throughout this book will show. *Winning the war* means removing the proteins that may harm us from our diets, finding antidotes to them or, in the best case, engineering them to remove drawbacks and allow us to take advantage of their useful functions. This is discussed in more detail in the first chapter, which deals with the molecular basis of toxicity and mechanisms by which proteins can become toxic. While common commercial names are occasionally cited, the reader should note that these may be changed by the manufacturer and as generic versions are introduced. Many of the treatments mentioned throughout the book may be obsolete within a few years, as clinical trials or their long-term use tests their efficacy and safety. They may be replaced by compounds with similar activity but less potential to cause adverse events. We are at a very exciting point in medicine, where broad spectrum therapies such as steroids and cytotoxic chemotherapies can be replaced by specific inhibitors of the proteins driving diseases. These may be targeted by small molecules or whole proteins. This also makes molecular and genetic testing more important, requiring much more complex data analysis. Proper diagnosis requires an open mind and ability to deal with complexities, something difficult to do in the short time a doctor is given to evaluate a patient.

None of the small molecules, proteins, vaccines, bacterial factors or antibodies described here should be taken as a treatment for any disease without consultation with a licensed medical authority.

Although written from a biochemist's standpoint, each chapter should be comprehensible for anyone who wants more understanding of proteins that underlie disease states and the complexities of therapies that use them. The unique features of each person: their genes, blood type, diet, environmental exposure, sensitivity to a given food or medicine, can all play a role in determining whether conditionally toxic proteins can help or harm them. For example, allergens and glutens are innocuous for one person but highly toxic to another. I hope that a mother of a peanut-allergic child, a newly diagnosed patient with gluten intolerance (or celiac disease), or someone suffering chronic headaches considering Botox injections, can use this book to better understand the proteins on the basis of selected diseases or their therapies. For those scientists engaged in this combat, or others who simply want to know more, illustrations are given and suggestions for further reading are listed at the end of each chapter. There are thousands of papers and whole books related to most of the proteins I have chosen to highlight. While I have given suggested further reading for those interested, the chapters are not intended to be comprehensive literature reviews.

My hope is that by understanding the dual nature of conditionally toxic proteins, we can better understand how to treat the diseases related to them. I wish all my readers the best of health and happiness.

Catherine Schein

1 How to Make a Toxin

Alle Dinge sind Gift, und nichts ist ohne Gift, allein die Dosis macht dass ein Ding kein Gift ist.

All things are poison, and nothing is without poison, the dosage alone makes a thing not a poison.

Paracelsus

OVERVIEW

1. Small molecules illustrate how even a single atom change can render a molecule virulent.
2. Even NSAIDs and other drugs sold over the counter can have risks (FDA-approved does not mean safe at any dose!).
3. The immune response to proteins that vary among individuals can cause immediate reaction to blood products or delayed hypersensitivity.
4. Essential mammalian proteins, such as insulin, interferon and cytokines, can induce a toxic immune response if overexpressed or used for the treatment of chronic disease.
5. Other proteins can misfold or form aggregates that are characteristic of diseases such as Huntington's, Alzheimer's or amyloidosis.
6. Modern therapeutics use antibodies to bind proteins or their receptors to block unwanted activities or stimulate T-cells.
7. Inhibitors of tumor necrosis factor (TNF) and interleukins that are key drivers in chronic diseases can provide more specific, alternative treatments to broad-spectrum steroid use.
8. Bacterial toxins can have many different activities and delivery modes that damage other microbes or mammalian tissues and cells.
9. For almost every toxin, there is a natural antitoxin or a use that takes advantage of its unique mechanism of action.
10. Understanding how toxins work can lead to beneficial uses.

DOI: 10.1201/9781003333319-1

TOXINS INTERFERE WITH METABOLIC PROCESSES OR DIRECTLY DAMAGE CELLS OR TISSUES

This clear definition requires a good deal of further information about where and how much of a toxin is present, as the examples throughout this book will show. Even the most innocuous molecule may be a toxin, at a high enough concentration or when present in the "wrong" place or time. The yin/yang concept can be applied to just about any molecule one can think of. As an extreme example, 28-year-old Jennifer Strange died in 2007 from drinking 2 gallons of water in 3 hours! However, one would not say water itself is a toxin: a larger individual might have consumed the same amount of water without a problem, and adding salts and metabolites might have warded off her body's electrolyte imbalance.

As examples throughout this book will show, even very small changes in molecular structure can generate virulent toxins.

Making a naturally produced molecule into a toxin may require only minimal change. While this book deals with conditionally toxic proteins, which are large molecules, it is useful to start with a very simple chemical example of how small changes can produce a toxin. Figure 1.1 shows one of the smallest naturally produced organic molecules, methane, which is a byproduct of normal digestion. Although the low methane levels in human emanations are unpleasant but not toxic, the large amounts of methane produced by cattle in enclosed quarters can indeed lead to toxic effects, both for the animals and the humans who care for them. In a more concentrated form, methane is a major component of natural gas, which is highly flammable.

While methane itself is thus not, strictly speaking, a toxin, small chemical modifications can produce very toxic molecules. Methane's basic structure is a carbon (C) atom surrounded by four hydrogen (H) atoms (Figure 1.1). However, converting just one of those H atoms to bromine (Br) or iodine (I), to make methyl bromide or methyl iodide (iodomethane), generates gaseous compounds that are much more toxic than methane itself. Apart from their effects on humans, these "halogenated" molecules

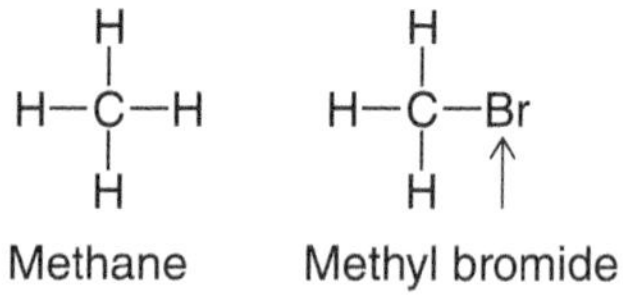

Methane Methyl bromide

FIGURE 1.1 From flatulence to fungicide: Methane (left), produced normally as a byproduct of digestion, can be converted to the toxic chemical methyl bromide (right), used for over 50 years to prevent the growth of weeds and fungus.

can damage the ozone layer. Methyl bromide has been used since the 1960s to fumigate soil before planting crops, as it prevents the growth of fungus and weeds. Recently, its use in agriculture has been largely eliminated in most countries. As it is inexpensive and effective, it may be used illegally in other countries. For example, a vacationing family from Delaware was poisoned by methyl bromide used to fumigate a villa where they stayed in the Virgin Islands. Although the exterminator who applied the chemical was brought to justice, the family was hospitalized with loss of sensation and paralysis, lasting several months.[1]

WHY WOULD ANYONE WANT TO MAKE A TOXIN IN THE FIRST PLACE?

There are several answers to this question. Methyl bromide was used as a fungicide to treat soil, on the premise that it would be degraded before finding its way into crops. Many different commercial herbicides and insecticides have been developed to specifically inhibit plant or insect essential processes, meaning they should have little or no activity affecting mammalian cells. Unfortunately, adverse events (AEs) of these chemicals are often only noticed (by bee species die-offs or increased cancer rates in humans exposed to them) after they have been used for a long time. As noted above, just because a drug or agricultural chemical has been approved does not mean it is "safe at any dose". Another illustration of this principle is ammonia-based fertilizers, relatively innocuous when distributed over a wide area, but explosive when stored in large amounts.

POTENTIALLY EXPLOSIVE AMMONIA-BASED FERTILIZERS HAVE EXPONENTIALLY INCREASED CROP YIELDS

The ability to synthesize vast quantities of ammonia, in industrial processes developed in the 1920s by Carl Bosch (Nobel Prize in Chemistry, 1931), is the bedrock of modern agriculture. Ammonia is a conditional toxin, a naturally produced molecule that must be continuously removed from blood (see the next chapter). Ammonia and fertilizers made from it are explosives. A ship loaded with ammonium nitrate caused the largest non-nuclear explosion in the history of the United States in the harbor of Texas City, Texas, in 1947. At least 580 people are estimated to have died in the disaster and more than 1000 injured, many with terrible burns. Another explosion of the same chemical, in West, Texas, in 2013, killed 15 people and destroyed or damaged hundreds of homes.

Safer methods to locally produce ammonia should eliminate the need to store and ship these dangerous fertilizers in large amounts. Additional efforts to introduce nitrogen into soil directly (though the application of nitrogen-fixing microorganisms to the roots of grain plants, for example) may even serve to eliminate ammonia-based chemicals completely in agriculture.

AMMONIA, CHLORINE AND "MUSTARD" GASES BECAME WEAPONS OF WAR

It is to the eternal shame of the human race that the methods used to produce and control gaseous ammonia were also used to introduce chlorine and later "mustard" gases to World War I battlefields. Mustard gas is a misnomer: there is no relation to mustard produced from the seeds of that plant and the molecule itself forms a sticky liquid. Aerosols of this were however easier to control than gaseous Chlorine, meaning they could be directed with horrible results against enemy troops.

Toxic Gas Weapons of War Stimulated the Search for Chemotherapeutic Agents

Doctors handling the survivors of the battlefield attacks realized that such a potent cell-killing agent might be the basis of cancer treatment, as fast-growing cells were most susceptible to destruction.[2] Chemists began a long search for compounds that could induce fast-growing cancer cells to lyse, while having less effect on slower replicating normal tissues. After many iterations, two alkylating agents, chlorambucil (Leukeran) and busulfan (Myleran) (Figure 1.2), became the first treatments for chronic lymphocytic and myeloid leukemias (CLL and CML). These simple and inexpensive drugs may still be used, although many more advanced (and less toxic) therapies are available today. Figure 1.2 also illustrates the fine line between helpful and harmful. Chemical tinkering with less admirable aims led to the deadly VX neurotoxin, which can kill at vanishingly low amounts if taken orally or absorbed through the skin.

Even Medications Sold Over the Counter (OTC) Can Be Toxic, Especially in the Context of Polypharmacy

There are dangerous chemicals lurking in our environment that, with proper handling, should be avoidable. What is more insidious are common, conditionally toxic medications. The majority of Americans have potentially inappropriate medications (PIMs), conditionally toxic drugs in their medicine chest, purchased freely in a supermarket or pharmacy.

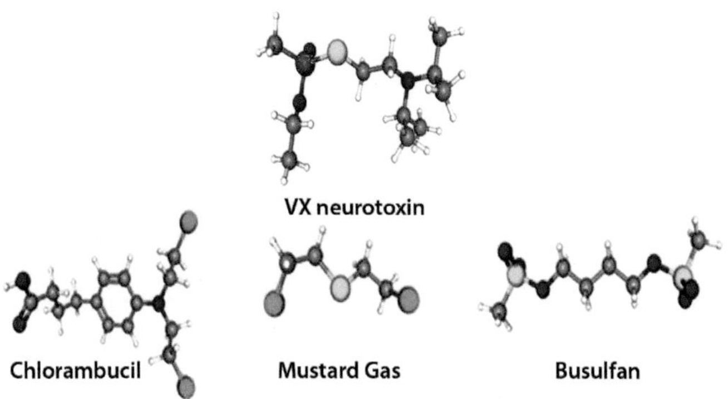

FIGURE 1.2 Small molecule chemical design leads to chemotherapy agents or deadly toxins: The framework of "mustard gas" (center) after many different iterations yielded the antileukemia drugs chlorambucil and busulfan. Modification with different goals gave the deadly VX neurotoxin (top). Structures are from PubChem with CPK element coloring: hydrogen = white, oxygen = red, chlorine = green, nitrogen = blue, carbon = gray, sulfur = yellow, phosphorus = orange.

For example, aspirin, a "non-steroidal anti-inflammatory drug" (NSAID), causes bleeding problems; taking high doses, such as those prescribed for severe arthritis, can induce dizziness, stomach problems and even convulsions. Doctors turned to recommending another NSAID, acetaminophen (paracetamol, Tylenol), as a safer alternative. Taken on its own at the recommended dose, this rapidly became the most widely used painkiller in the world. A spoonful of pink liquid Tylenol can bring a child's fever down quickly. However, cheap, generic acetaminophen may also be added to cough syrups, headache tablets and other complex medications. Few realize how close the toxic dose of this seemingly innocuous drug lies to its therapeutic one, meaning that taking multiple tablets, or combining with other medications or alcohol, can lead to nausea, vomiting and, most dangerously, fulminant liver failure. Thousands of hospital visits and hundreds of fatalities annually are attributed to acetaminophen poisoning in the US alone.

The toxicity of acetaminophen lies in its metabolite (Figure 1.3), NAPQI (or NAPBQI, *N*-acetyl-*p*-benzoquinone imine), binding to cysteines and deactivating proteins. At low concentrations, NAPQI is conjugated to a simple molecule, glutathione, which prevents it from protein binding. But normal glutathione concentrations may not be sufficient as more acetaminophen enters the blood stream and eventually the liver. NAPQI accumulates in a free form, binds to proteins and nucleic acids in cells and causes severe toxicity.[3] Fortunately, if the problem is seen early enough, infusion

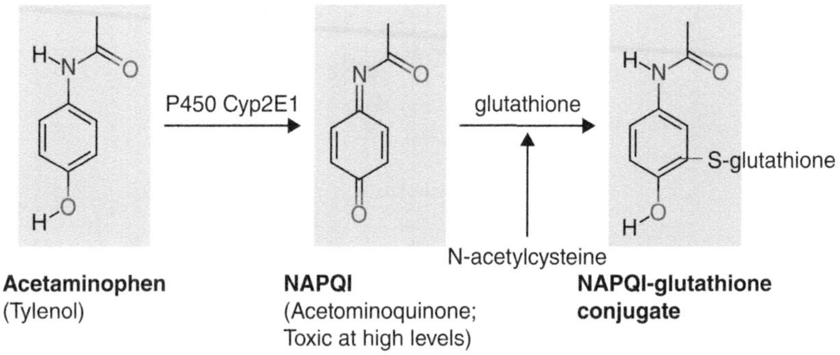

P450 Cyp2E1

glutathione

S-glutathione

N-acetylcysteine

Acetaminophen
(Tylenol)

NAPQI
(Acetominoquinone;
Toxic at high levels)

NAPQI-glutathione conjugate

FIGURE 1.3 **Toxicity related to acetaminophen** is primarily due to metabolites such as NAPQI, which can react with cysteines to deactivate proteins. When sufficient glutathione is present, it forms a conjugate with the metabolite, which prevents protein binding. In overdose situations, *N*-acetyl cysteine (NAC) can be infused to increase glutathione concentration.

of large amounts of *N*-acetyl cysteine (NAC) (Figure 1.3) can increase the glutathione pool, allowing clearance of NAPQI and normal liver function to resume. Although one could drink NAC, it smells like rotten eggs!

Steps are being taken more recently to cut down on accidental overdoses and warn patients of the danger of combining acetaminophen with other drugs.[4]

TOXINS AND TOXIC PROTEINS EVOLVED AS DEFENSIVE WEAPONS OF MICROBES AND PLANTS

Small molecule and protein toxins evolved in both uni- and multicellular organisms to protect them against surrounding pathogens. Successful toxins are often produced by many different bacteria or plants. For example, the shells of cashew nuts and pistachios contain urushiol, a toxin best known for causing the rash of poison ivy. These nuts must be roasted or boiled and their shells removed before the edible portion can be marketed.[5] Mango skins also contain urushiol and can cause "type IV" (non-IgE mediated) reactions to those previously sensitized to poison ivy.[6]

Moving to protein toxins, those of bacteria, plants and animals[7] have evolved over billions of years, honed in ongoing battles that only incidentally involve mammals. As described in more detail in Chapter 8 (Converting Bacterial Toxins to Human Therapeutics) of this book, microbes are continually at war with each other, for survival and eventually dominance. Natural microbial communities consist of complex mixtures of bacteria, viruses, fungi and algae, with each organism fighting against the others for dominance. An organism survives by secreting toxins and antibiotics (such as penicillin, produced by a fruit fungus), and by evolving internal enzymes and exporters that cause other compounds that penetrate its own cell wall to be secreted into the surrounding medium. Potent toxins can often be produced by seemingly unrelated plants or organisms. A similar venom is produced by many different wasp species and even fire ants.

The first reason for producing toxins and superantigens is to enhance the survival of the producing organism in a complex environment. Thus, it makes sense that many toxins are only produced when bacteria are in the late stage of growth, as they are reaching senescence. Spores of the soil bacteria *Bacillus anthracis* (the causative agent of anthrax) are coated with toxins that are destined to kill other bacteria and fungi, protecting the non-growing but still viable spores. When cows or sheep eat the grass infected with these spores, they can regenerate into rapidly growing bacterial cells that reach other areas of the digestive system, where their toxins provide them an advantage over competing microbes. The spores also can cause black sores on the skin of wool and leather workers who come in contact with them. Eating or breathing in spores can lead to painful and sometimes deadly disease unless antibiotic treatment (to kill the regenerating bacteria) is started early.

OUR IMMUNE SYSTEM, EVOLVED TO FIGHT DEADLY PLAGUES, CAN OVERREACT TO INNOCUOUS STIMULI

Our own genes and proteins have also been affected by the battle to survive, including occasional contact with anthrax and other bacteria. Up until the 20th century, innate immunity was our major weapon against microbial pathogens. Our immune system, and as we will see in Chapter 2, even our blood cells were developed and selected for use during our own wars against these invaders. While many of the small battles that shaped our immune responses would be hard to follow genetically at this time, one major one stands out. In the middle of the 14th century, the Black Plague, caused by the bacterium *Yersinia pestis*, raged through Europe, killing about half the population of England (Figure 1.4). The genome obtained from bodies of victims in both England and Denmark showed how the plague selected for immune system genes in survivors. Alleles for the ERAP2 gene (endoplasmic reticulum aminopeptidase 2) in particular stood out in the analysis as having become more common in survivors of the disease.[8] Macrophage ERAPs chop up the proteins of invaders to prepare the resulting peptides for presentation to activated T-cells, stimulating them to release cytokines and eventually determining the specificity of antibodies made by B-cells. All this makes sense of course: eventually those who could better generate an immune response to *Y. pestis* had the

FIGURE 1.4 **The Black Plague,** which killed half the population of England in the middle of the 14th century, may have selected for survivors with a more reactive immune system, setting the stage for some autoimmune diseases and sensitization to innocuous proteins.

best chance of surviving the plague and passing their genes on to their children.

But the ERAPs were already well known to medical detectives before this study found possible natural selection in plague survivors. Overexpression of ERAP proteins has been associated with immune-related syndromes, and ERAP2 has been particularly implicated in Crohn's disease. This led to the conclusion that, in selecting for a more powerful response to bacterial attack, the immune system also evolved to overreact to other stimuli.

Immune reactions can be deadly, the best known being responses that cause difficulty in blood transfusions. Chapter 2, "Blood and Gas", looks at the antigens that differ among populations, making transfusions and tissue transplantations so difficult to standardize. Selecting the right donor for a given patient (receiver) is important to prevent potentially lethal immune reactions, but determining a person's blood type can be complicated. Besides the major blood groups, the genetically encoded lack or presence of protein antigens that stimulate the immune system differs from person to person.

While the immune response to blood antigens is reasonably well understood, overreactions to foods or pollen that occur in only some individuals are not. Foods containing wheat (Chapter 3) or peanuts (Chapter 4), which are consumed by most of the population, can be toxic to individuals if they have been sensitized to these proteins (Figure 1.5). Chapter 3, "Minding the Ps and Qs of Gluten" describes what makes these conditionally toxic

FIGURE 1.5 Allergenic proteins are found in many foods. While most of the population consumes these foods with no ill effects, they are toxic for those with sensitivities.

proteins so odd. Most people eat gluten, in products ranging from bread to sausages, with no ill effects. However, for individuals with celiac disease (CeD), eating gluten causes symptoms ranging from headaches to skin rash to chronic diarrhea. Understanding the immune response to peptides containing high concentrations of the amino acids proline (P) and glutamine (Q) has enabled assays that can aid in diagnosing CeD even without a biopsy. Unfortunately, elimination diets (discussed in the Introduction to Chapters 3 and 4), which can be difficult to maintain, may be the only treatment for food-related sensitivities.

Data suggests that the incidence of autoimmune and allergic diseases is increasing. At the same time, we have received an arsenal of unique weapons: vaccines, and when these fail, antibiotics and antivirals, meaning we no longer fear extinction from the infectious diseases that killed so many in generations past. Perhaps we can use the clues in allergens and other stimuli to understand the potential linkage between these changes in the human condition. This may lead to better control of our deleterious immune responses, directing their activities against our true chemical or protein adversaries. In the meantime, we must use these gifts wisely, so they can be useful against another plague, caused, for example, by an antibiotic-resistant bacterium or a rapidly spreading viral mutant.

WHEN ESSENTIAL PROTEINS BECOME TOXIC

Just as a headache remedy can lead to liver failure, essential proteins produced by the human body, such as insulin or interferons (IFN), can be conditionally toxic, depending on the amount and location. Too much of a good protein, even one produced by the human body itself, can be toxic. Overdosing insulin, the most vital protein in the body, can cause cells to take up blood sugar too rapidly, thus diminishing sugar levels in serum to dangerously low levels. If glucose levels in the blood drop so low that the brain no longer functions properly, this can lead to shock, coma, and even death.

There are many mechanisms for a protein to become toxic. A protein can mutate, lose an essential function, gain a deleterious function (such as interfering with processes such as autophagy) or simply accumulate age-related changes in its structure that lead to aggregation. Some human proteins directly cause cells to die, such as the aptly named tumor necrosis factor (TNF). Although inducing cell death and increasing the amounts of other proteins (such as interleukin-1β (IL-1β) that cause fever would seem to be negative effects, TNF itself, or inducers of it, may be a useful anticancer agent. Other proteins, such as IFN and interleukin-2 (IL-2), are essential for correct immune system function. However, their presence in the body is very tightly controlled. IL-2, for example, may be very useful in cancer therapy, as it increases "NK-cells" (natural killers). However, even short-term therapy with IL-2 has caused human deaths. Long-term therapy with IFN, for example to treat chronic hepatitis C infections, can possibly lead to another dreaded chronic disease: lupus erythematosus. Thus, essential, valuable proteins can be conditional toxins, depending on concentration and exposure time.

SHAPE-SHIFTING AND AGGREGATING PROTEINS

Even normal mammalian proteins and their breakdown products can become toxins by changing their shape and interaction with other protein molecules. Figure 1.1 shows how methane can be converted to a toxin, just by changing one hydrogen atom. Changing a bond orientation in a small molecule can make the difference between active and inactive drugs. As discussed in Chapter 7 (Aggregation and Solubility: Making Milk from Cheese), turning a bond between glucose moieties in one direction makes water-soluble, digestible starch. Reversing that direction makes insoluble cellobiose, the basic polymer in grasses and wood.

In a similar fashion, amino acids in proteins or whole protein chains may change their orientation to each other, to form insoluble aggregates. Proteins can be induced to aggregate by small changes in structure[9,10] or temperature.[11] Aggregation of proteins is the basis of cheese making, blood clotting and fertilization. Chapter 7, "Aggregation and Solubility", deals with the positive and negative consequences of protein aggregation. Just as one cannot easily get milk back once it has formed cheese, or flour once it has been made into a dough, one cannot usually make an aggregated protein return to its native state.

Aggregates of proteins characterize many neurological diseases. These include several diseases referred to by their discoverer's name, Alzheimer's, Parkinson's and Huntington's, as well as those named for the causative proteins, such as lambda light chain disease (LLCD). Scientists have struggled for years with how to control protein solubility and prevent aggregation in these diseases,[12,13] research which is finally yielding treatments. Some protein aggregation diseases, such as Huntington's and ataxias, are characterized by the expansion of genes encoding long stretches of a single amino acid, glutamine, which can lead to protein aggregates in cells and gradually worsening dementia. Glutamine-rich repeats are also found in glutens and have been suggested to play a role in the neurologic complications that frequently accompany intestinal issues in those with CeD (Chapter 3).

Other proteins may lead to devastating diseases, such as Alzheimer's and Parkinson's, by aggregating to form "clumps" in brain cells. In addition to losing their ability to perform their normal functions, the aggregated protein "plaques" in the brain interfere with metabolism. Eventually, these aggregates rob an individual of the ability to remember and distort their personality. Thus, unique solutions are being tested for these diseases, including preventing the production of proteins that form the aggregates, or antibodies that bind to the aggregates and help clear them from cells.[14,15]

Figure 1.6 shows an example, based on the Aβ peptide that aggregates and accumulates in the brain cells of people with Alzheimer's disease. In normal cells, aggregated or "unfolded" proteins can be disposed of. However, protein aggregates are resistant to proteolysis, even with some of the most active proteases. Antibodies that can target and break apart Aβ plaques are available, but their effect on cognition may not be clear for reasons discussed in Chapter 7.

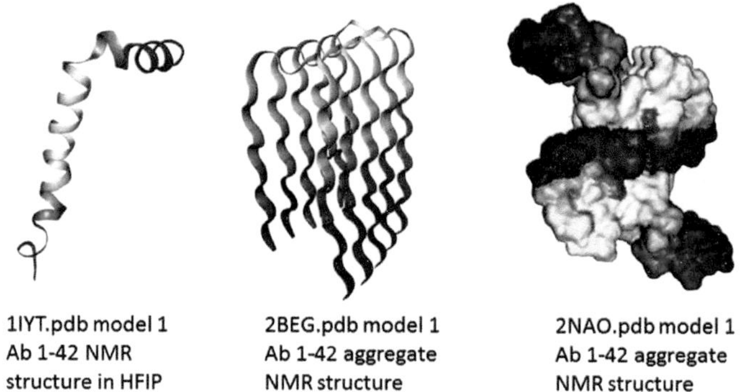

FIGURE 1.6 Shape-shifting, aggregating protein. The 42 amino acid Aβ peptide forms aggregates in brain cells, seen as plaques in Alzheimer's disease brain. The soluble form of the peptide (far left) forms an α-helix in a solvent also known to favor helix conformation. When transferred to water, it forms primarily β-stranded insoluble fibrils, with the N-termini more disordered, as structures based on data from solid state NMR (middle, fibril backbone) and space filling from solid state NMR coupled with other visualization methods[16] (far right) show. The depictions are based on the indicated files in the Protein Database (PDB).

TOXINS AND ANTITOXINS

For each potentially toxic protein, there is typically a natural antidote. An EMS person, faced with a diabetic who has accidentally taken too much insulin, may inject him with glucagon, a natural hormone protein that has the opposite effect of insulin. While insulin causes cells to take up glucose, glucagon stimulates them to release it. Similarly, the "fever cytokine" IL-1β, a player in cytokine storm (Chapter 6), is one of a family of related proteins, which also includes a natural inhibitor. Although mice engineered to not produce IL-1β live normal lives, those lacking its inhibitor are non-viable. As Chapter 6 discusses, there are pharmaceutical ways to treat diseases caused by the over-activity of some proteins, even when no natural inhibitor is known. In diseases such as psoriasis, or rheumatoid arthritis, inhibiting the activity of an overproduced protein such as TNF can be a path to therapy (Chapter 6).

Another path is to exploit the specificity of the immune response to produce an antibody that binds specifically to a protein and stops it from exerting its activities (as described for TNF family proteins in Chapter 6, and bacterial antitoxins in Chapter 8). Anifrolumab, an antibody that blocks the cell receptor for IFN, has recently been approved for the treatment of severe lupus.[17]

WHY DOES ITS VENOM NOT KILL THE SNAKE?

Anti-venoms, specific antibodies usually made in horses, goats or other large animals, that bind to venom proteins and prevent them from doing further tissue damage, have long been the basis of treatment for snake bites. To get the antibodies, one uses the Paracelsus adage quoted, that the amount (dose) makes the toxin. Injecting small amounts of snake venom does not make a horse sick but does stimulate its immune system to make antibodies against the venom proteins that are perceived, correctly, as hostile invaders. After a few weeks, with boosting from additional small quantities of the toxin, mature and specific antibodies can be purified from the horse's blood.

This leads to the question: why does the toxin not kill the snake, which is much smaller than a horse? Again, nature has many ways of keeping toxins from harming their producers. In the simplest form, snake venom is produced and stored in a separate gland and released only when the unsuspecting prey is attacked. Alternatively, the producer may have mutations in its own proteins that make them resist the effects of the toxin.

Anti-venoms may save lives, but they are not without problems.[18] First, the anti-venom antibodies are typically very expensive and need to be administered intravenously, meaning that they cannot be carried routinely in the backpack of the outdoorsmen who are likely to be bitten. Even if they were affordable, the anti-venom needs to be kept from extreme heat and properly reconstituted if freeze-dried. Small, stable molecules to inhibit snake venoms are lacking.

BACTERIA ARE GREAT PROTEIN BIOENGINEERS

Bacterial toxins are often related to and probably derived via mutation of internal enzymes. To provide weapons in its war with other microbes, mutations may convert a copy of an essential protein to a toxin, which becomes fixed in its genome when only the fittest survive. Such "natural selection" favors the survival of those secreting toxins that can be injected into the cytoplasm of their targeted cells. In the anthrax case, a protease called lethal factor and a version of adenylyl cyclase called edema factor[19–21] are produced along with a protein that facilitates their entry into the cytoplasm of other bacteria. Even completely unrelated bacteria may acquire such toxins from the original producer, by taking up small pieces of DNA containing the genes for toxins. For example, many secrete adenylyl cyclase toxins that directly convert ATP to cAMP. Alternatively, bacteria such as cholera produce toxins that work intracellularly to induce cellular enzymes to increase their output of cAMP. High tissue levels of cAMP can work very quickly to induce edema in the host. The effect of cholera toxins on the human body has been shown by many epidemics throughout the centuries. Writers have described how cholera infection can turn a normal human into "a cold gray corpse" in a few hours, if the devastating fluid loss is not dealt with promptly. However, it is still not clear how the elevated cAMP levels help the bacteria themselves. One possibility is that the cAMP signaling wave paves the way for the rapid growth of bacteria, which can double in number every 20–30 minutes if not stopped by an active immune response.

Chapter 8 deals with bacterial toxins and their natural and synthetic inhibitors. It also shows how even the most lethal toxins can be medically useful.

PATHOGENIC BACTERIA USE TOXINS TO ADAPT TO DISCRETE AREAS IN MAMMALS

In some form, relatives of even the most potent pathogens occupy many "privileged niches" in the body, such as the intestinal track, semen, nostrils and outer layers of the skin, where they do little harm. These resident microbes may have very thick walls that can protect them against even deadly animal venoms.[22] If allowed to enter less protected areas, bacteria secrete a series of protease toxins that degrade tissues, providing them with needed nutrients for their rapid growth. Almost all pathogenic bacteria have some version of these toxins. Once they have obtained access to the inner layers of the skin, the kidneys or the blood, they can use their cell-penetrating machinery to inject enzymes such as adenylyl cyclases into mammalian cells to cause further damage. Mammalian cell walls are relatively easy to penetrate compared to the multilayered cell walls of bacteria. Continued production of cAMP may also slow cell growth and make the environment even more amenable to bacterial survival. Tuberculosis bacteria have more than ten adenylyl cyclases to increase cAMP in tissues![23] In addition to their powerful offensive toxins, most clinically important pathogens have acquired defensive weapons for surviving within mammalian tissues. These include enzymes to degrade or export antibiotics, protease inhibitors to block serum proteases and other tools to evade the mammalian immune system.

ANTIBIOTICS MAY LEAD TO THE SPREAD OF "SUPERPATHOGENS"

Modern methods of dealing with infections may further the growth of organisms equipped with an array of tissue-damaging toxins and resistance genes that will defeat most antibiotics. For example, cystic fibrosis patients and others with severe immunosuppression are discouraged from meeting each other in person, to prevent them from passing antibiotic-resistant pathogens from one to another. Diabetics must also fear flesh-eating bacteria, superpathogens which can enter the body through even small scratches. If treatment is delayed, infection can lead to necrotizing fasciitis, total organ failure and death within a day.

What is perhaps surprising is that many toxic proteins, including necrotizing factors produced by human cells (Chapter 6) and neurotoxins of pathogenic bacteria (Chapter 8), are potentially valuable therapies for some diseases. Chapter 8, "Converting Bacterial Toxins to Human Therapeutics", highlights just a few of the medical uses of toxins produced by bacteria and other pathogens. Even the famous botulinum neurotoxin, which has caused many deaths from eating improperly conserved food, has numerous cosmetic and medicinal uses. Understanding how these proteins injure cells can allow them to be repurposed for human therapeutics, making even the deadliest of them "conditionally toxic proteins".

Beneficial use requires knowing the mechanism of toxicity. Sometimes this is obvious, while for other proteins one should know their function in their native environment, and what stimulates an immune response. Then, chemically, we must know how to change the protein to mitigate toxicity while enhancing the activity that will make them useful to us.

MODERN PROTEIN SOURCES MAY BE DESIGNED TO AVOID ALLERGENS OR TOXINS

Post-translational modifications, or contamination from other compounds in the sources the protein is isolated from, may also render a harmless protein toxic. The use of "single cell proteins" (SCP) was introduced in the 1970s as a way to solve the world's nutrition problems by producing microbes for food use from very low-cost substrates or better yet, those that would cost a lot to get rid of, such as diluted suspensions of oil (e.g. the wash water of tankers), sawdust or water from potato processing.[24] Vast amounts of yeast (normally found in the diet as a leavening for bread) or edible fungi (delectable mushrooms and truffles are fruiting bodies of fungi) could be made. Although most problems in scale-up were solved from an engineering standpoint, food chemists ran into a major problem: the yeast or fungi took up the taste and contaminants of any substrate they were grown on, thus rendering them *conditionally toxic*. Covering the bad taste with lots of sugar and fat rather defeated the original purpose of obtaining high-protein foods. The processes lost favor when it became clear that the "waste streams" lent themselves better to ethanol and production of other chemicals.

But these early experiments showed that it was possible, using food-grade ingredients, to grow large quantities of delectable fungi that could be a useful meat alternative, low in saturated fat and high in essential amino acids. Modern artificial meats start with vegetable sources that are already suitable for human consumption. They may still cause problems for those with soy, pea or wheat allergies. Thus, there is a continuous search for hypoallergenic proteins that can be used to construct more textured products. There is now a growing and potentially very profitable market for these products as they steadily improve in taste and variety. If we have a better idea of possibly problematic areas, engineered proteins can be designed to be hypoallergenic and easily digested, free of "food maps" or gluten.

CONCLUSIONS

As these examples show, there are many ways to make a toxin or for a protein to be toxic. If we can control the activity of these toxins and make them do what we want them to, where they are needed, they can be valuable therapeutics. Most fascinating is how even the most dangerous molecules can be used for beneficial purposes. Given time, this may allow these proteins to save as many lives as their pathogenic predecessors have taken. Many novel treatments have been derived to control toxins. New ways of producing proteins may allow one to produce foods that are safe for all to eat and provide cheap nutrition for a hungry world.

The next chapters will cover proteins in blood or food that trigger immune system reactions. Although elimination diets are very hard to maintain, the major "treatment" for CeD and allergies is currently avoidance of a specific triggering source. Newer treatments may come from a better understanding of how to control immune responses.

REFERENCES

1. Bies J. Terminix manager sentenced to 12 months in prison for Delaware family's poisoning. *Delaware Online*. 1/22/2019. https://www.delawareonline.com/story/news/2019/01/22/terminix-manager-sentenced-12-months-prison-familys-poisoning/2643153002/
2. Schein CH. Repurposing approved drugs for cancer therapy. Br Med Bull. 2021;137(1):13–27. doi: 10.1093/bmb/ldaa045. PubMed PMID: 33517358; PMCID: PMC7929227.
3. McGill MR, Hinson JA. The development and hepatotoxicity of acetaminophen: reviewing over a century of progress. Drug Metab Rev. 2020;52(4):472–500. Epub 20201014. doi: 10.1080/03602532.2020.1832112. PubMed PMID: 33103516; PMCID: PMC8427730.
4. Schilling A, Corey R, Leonard M, Eghtesad B. Acetaminophen: old drug, new warnings. Cleve Clin J Med. 2010;77(1):19–27. doi: 10.3949/ccjm.77a.09084. PubMed PMID: 20048026.

5. CDC. Dermatitis associated with cashew nut consumption – Pennsylvania. MMWR Weekly. 1983;32(9):129–30.
6. Yoo MJ, Carius BM. Mango dermatitis after urushiol sensitization. Clin Pract Cases Emerg Med. 2019;3(4):361–3. Epub 20190930. doi: 10.5811/cpcem.2019.6.43196. PubMed PMID: 31763588; PMCID: PMC6861053.
7. Negi SS, Schein CH, Ladics GS, Mirsky H, Chang P, Rascle JB, Kough J, Sterck L, Papineni S, Jez JM, Pereira Mouries L, Braun W. Functional classification of protein toxins as a basis for bioinformatic screening. Sci Rep. 2017;7(1):13940. Epub 20171024. doi: 10.1038/s41598-017-13957-1. PubMed PMID: 29066768; PMCID: PMC5655178.
8. Klunk J, Vilgalys TP, Demeure CE, Cheng X, Shiratori M, Madej J, Beau R, Elli D, Patino MI, Redfern R, DeWitte SN, Gamble JA, Boldsen JL, Carmichael A, Varlik N, Eaton K, Grenier JC, Golding GB, Devault A, Rouillard JM, Yotova V, Sindeaux R, Ye CJ, Bikaran M, Dumaine A, Brinkworth JF, Missiakas D, Rouleau GA, Steinrucken M, Pizarro-Cerda J, Poinar HN, Barreiro LB. Evolution of immune genes is associated with the Black Death. Nature. 2022. Epub 20221019. doi: 10.1038/s41586-022-05349-x. PubMed PMID: 36261521; PMCID: PMC9580435.
9. Schein CH. Solubility as a function of protein structure and solvent components. Bio/Technology. 1990;8(4):308–17. doi: 10.1038/nbt0490-308.
10. Schein CH. Production of soluble recombinant proteins in bacteria. Bio/Technology. 1989;7(11):1141–9. doi: 10.1038/nbt1189-1141.
11. Schein CH, Noteborn MHM. Formation of soluble recombinant proteins in *Escherichia coli* is favored by lower growth temperature. Bio/Technology. 1988;6(3):291–4. doi: 10.1038/nbt0388-291.
12. Bruce NJ, Chen D, Dastidar SG, Marks GE, Schein CH, Bryce RA. Molecular dynamics simulations of Abeta fibril interactions with beta-sheet breaker peptides. Peptides. 2010;31(11):2100–8. Epub 20100804. doi: 10.1016/j.peptides.2010.07.015. PubMed PMID: 20691234.
13. Chen D, Martin ZS, Soto C, Schein CH. Computational selection of inhibitors of Abeta aggregation and neuronal toxicity. Bioorg Med Chem. 2009;17(14):5189–97. Epub 20090527. doi: 10.1016/j.bmc.2009.05.047. PubMed PMID: 19540126; PMCID: PMC2743868.
14. Chapleau M, Iaccarino L, Soleimani-Meigooni D, Rabinovici GD. The role of amyloid PET in imaging neurodegenerative disorders: a review. J Nucl Med. 2022;63(Suppl 1):13S–9S. doi: 10.2967/jnumed.121.263195. PubMed PMID: 35649652.
15. Plowey ED, Bussiere T, Rajagovindan R, Sebalusky J, Hamann S, von Hehn C, Castrillo-Viguera C, Sandrock A, Budd Haeberlein S, van Dyck CH, Huttner A. Alzheimer disease neuropathology in a patient previously treated with aducanumab. Acta Neuropathol. 2022. Epub 20220517. doi: 10.1007/s00401-022-02433-4. PubMed PMID: 35581440.
16. Wälti MA, Ravotti F, Arai H, Glabe CG, Wall JS, Böckmann A, Güntert P, Meier BH, RiekR. Atomic-resolution structure of a disease-relevant Aβ(1–42)amyloid fibril. PNAS. 2016;113(34):E4976–84. doi: 10.1073/pnas.160074911.

17. Morand EF, Furie RA, Bruce IN, Vital EM, Dall'Era M, Maho E, Pineda L, Tummala R. Efficacy of anifrolumab across organ domains in patients with moderate-to-severe systemic lupus erythematosus: a post-hoc analysis of pooled data from the TULIP-1 and TULIP-2 trials. The Lancet Rheumatology. 2022;4(4):e282–e92. doi: https://doi.org/10.1016/S2665-9913(21)00317-9.
18. Gutierrez JM, Fan HW, Silvera CL, Angulo Y. Stability, distribution and use of antivenoms for snakebite envenomation in Latin America: report of a workshop. Toxicon. 2009;53(6):625–30. doi: 10.1016/j.toxicon.2009.01.020. PubMed PMID: 19673076.
19. Schein CH, Chen D, Ma L, Kanalas JJ, Gao J, Jimenez ME, Sower LE, Walter MA, Gilbertson SR, Peterson JW. Pharmacophore selection and redesign of non-nucleotide inhibitors of anthrax edema factor. Toxins (Basel). 2012;4(11):1288–300. Epub 20121108. doi: 10.3390/toxins4111288. PubMed PMID: 23202316; PMCID: PMC3509708.
20. Chen D, Ma L, Kanalas JJ, Gao J, Pawlik J, Jimenez ME, Walter MA, Peterson JW, Gilbertson SR, Schein CH. Structure-based redesign of an edema toxin inhibitor. Bioorg Med Chem. 2012;20(1):368–76. Epub 20111116. doi: 10.1016/j.bmc.2011.10.091. PubMed PMID: 22154558; PMCID: PMC3251925.
21. Chen D, Misra M, Sower L, Peterson JW, Kellogg GE, Schein CH. Novel inhibitors of anthrax edema factor. Bioorg Med Chem. 2008;16(15):7225–33. Epub 20080628. doi: 10.1016/j.bmc.2008.06.036. PubMed PMID: 18620864; PMCID: PMC2678011.
22. Esmaeilishirazifard E, Usher L, Trim C, Denise H, Sangal V, Tyson GH, Barlow A, Redway KF, Taylor JD, Kremyda-Vlachou M, Davies S, Loftus TD, Lock MMG, Wright K, Dalby A, Snyder LAS, Wuster W, Trim S, Moschos SA. Bacterial adaptation to venom in snakes and Arachnida. Microbiol Spectr. 2022:e0240821. Epub 20220523. doi: 10.1128/spectrum.02408-21. PubMed PMID: 35604233.
23. Johnson RM, McDonough KA. Cyclic nucleotide signaling in *Mycobacterium tuberculosis*: an expanding repertoire. Pathog Dis. 2018;76(5). doi: 10.1093/femspd/fty048. PubMed PMID: 29905867; PMCID: PMC6693379.
24. Durkin A, Guo M, Wuertz S, Stuckey DC. Resource recovery from food-processing wastewaters in a circular economy: a methodology for the future. Curr Opin Biotechnol. 2022;76:102735. Epub 20220526. doi: 10.1016/j.copbio.2022.102735. PubMed PMID: 35644060.

2 Blood and Gas

OVERVIEW

1. Conditionally toxic proteins on blood cells vary across populations; the donor and receiver must be accurately matched to avoid severe immunologic reactions.
2. "Natural" antibodies causing reactions to "the wrong blood" may have been induced by similar antigens on external sources, such as animal viruses, during the first year of life.
3. Genes encoding antigens with similar phenotypes exhibit considerable polymorphism.
4. There are well over 1000 blood antigens, from about ~346 groups that vary among populations and individuals.
5. Matching of multiple cell membrane proteins is particularly important to prevent rejection of tissue transplants and isoimmunization of those receiving repeated transfusions.
6. Immunologically problematic proteins, including the Rh proteins, can have important physiological roles in aiding gas transport, maintaining blood pH or red blood cell shape.
7. Expression of sickle cell hemoglobin and Duffy antigens may be selected for or against in malaria-prone regions.

KARL LANDSTEINER AND ABO BLOOD TYPING

Perhaps the best known example of conditionally toxic proteins comes from attempts over centuries to save wounded soldiers with the blood of their compatriots.[1] Early (pre-20th century) experiences with transfusing blood from an animal or even one person to another were spectacular failures due to immune reactions against the donor blood. The first demonstrations that blood transfer was even possible were when parents were connected to their children (by sewing their veins together, a process called anastomosis). We now know, thanks to the work of Karl Landsteiner and his group in Austria at the beginning of the 20th century, that humans can only accept the blood of another person if the antigens and proteins on the surface of red blood cells (RBCs) match as closely as possible. As the

DOI: 10.1201/9781003333319-2

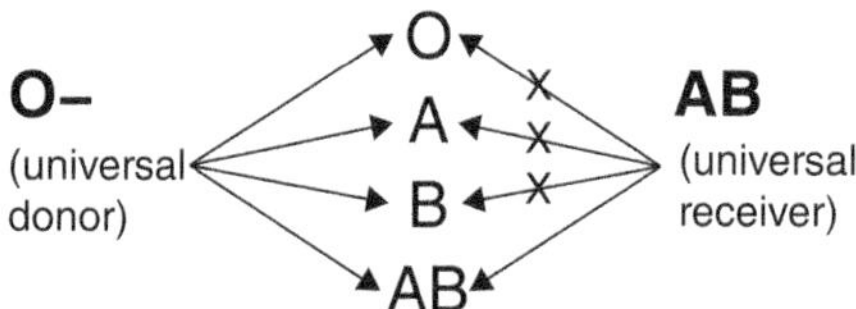

Receiver ABO:	Has antibodies to:			
	0	A	B	AB
O Type	–	+	+	+
A Type	–	–	+	+
B Type	–	+	–	+
AB Type	–	–	–	–

FIGURE 2.1 Agglutination of blood cells due to "natural antibodies" (+ in the table) occurs when blood is mixed with a different type. Type O^- cells express neither the A^- or B^- antigen and can be given to any person (universal donor) but O-type individuals have antibodies that develop in the first years of life against both antigens (bottom table). Only AB^+ individuals have no natural antibodies and can be universal receivers of all blood types. However, type AB blood can only be given to an AB person.

enzymes that dictate blood type are inherited, a child and father are more likely to be a suitable match than brothers at arms would be.

By 1910, Landsteiner and others had defined three blood groups, based on how sera from one individual caused "foreign" blood to clump. Landsteiner decided there were groups of people who could give their blood to one another (called the ABO system for short, Figure 2.1). Thus, A-type people could only exchange blood with others of the A type and B with other Bs. Type A blood would not aggregate if mixed with the blood of another A type. Type B blood would aggregate when mixed with A type but not B type. The group also recognized a third, very common "O" type blood, that was not A or B. O-type blood could be given to all individuals, regardless of blood type. This does not work reciprocally: an O-type person must receive only O-type blood.

We now know that this selective aggregation was primarily due to the different pattern of amino-sugar groups marking the proteins on the surface of an individual's RBCs (Figure 2.2). The difference is that A has *N*-acetylgalactosamine (blue in Figure 2.2) added to the end of a polysaccharide on the RBCs, while B has only galactose.[2] O-type RBCs have neither. This seems like a small change, but the immune reaction is so sensitive, it detects the difference of one sugar group, or as is the case between types A and B, a single N-acetylamine side chain.

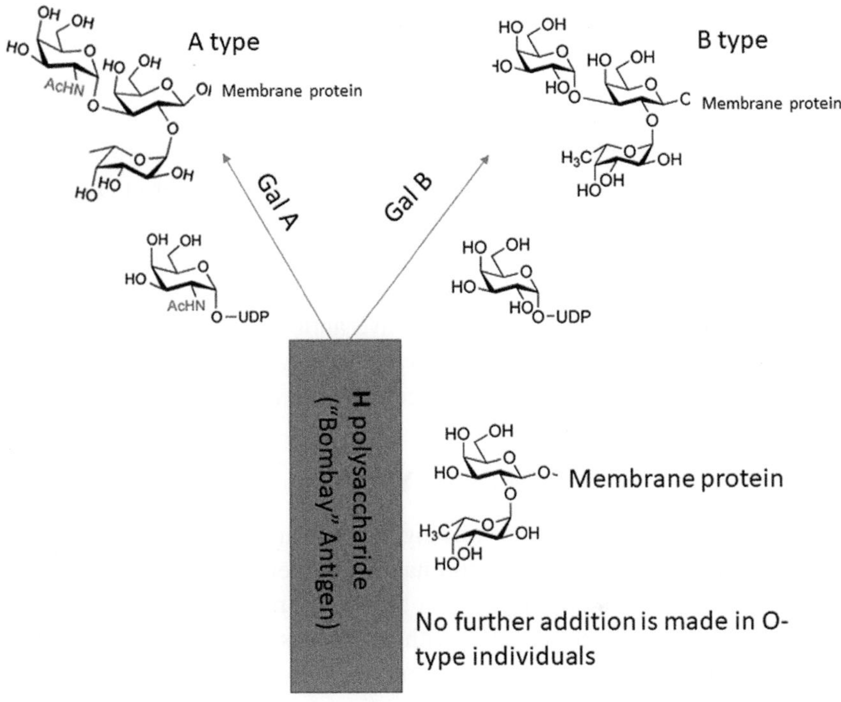

FIGURE 2.2 The A and B antigens differ only in the presence of an N-acetyl amine side group on the galactose transferred to the H⁻ ("Bombay") antigen from a UDP conjugate. The specificity of the addition is determined by the genetically encoded enzymes, galactosyltransferase type A or B (GALA or GALB). These enzymes are not active in Type O individuals; mutations and other factors may determine how much of each antigen is present on cells.

The AB blood type, where RBCs produce both A- and B-type antigens, is rare: <4% of Americans, compared to about 40% each of A and O blood types and about 10% with type B. Landsteiner's group did not even recognize it as a separate grouping when defining the system. He defined it initially as "null" as AB blood did not aggregate when mixed with Type A or B blood types.

There is another blood group that is even rarer than AB: some people (in the United States, about 1 in a million) lack the bottom part of the polysaccharide side chain, a disaccharide which presents the A and B antigens to the immune system. This area, called the "H" or Bombay antigen,[3] is lacking in a genotype found sporadically throughout the world but first identified in India, where it occurs in about 1 in 10,000 people. Individuals lacking the H antigen can only be treated with H^- blood,[4] which is not typically available in most blood banks in the United States.

As is clear from the frequency of occurrence of the H^- phenotype above, and in examples seen later in this chapter, blood groups are genetically determined and vary from one population to another. Americans are Type O (~44%) or A (~42%); African and Asian Americans have a two to three fold higher likelihood of being Type B than Caucasians or Latinos. Only about 4% of Americans have Type AB; this type is even rarer in Germany. Landsteiner, as noted above, described AB originally as "of no particular type" since it did not agglutinate when mixed with any of the other three blood types. We now know this is just the opposite. AB people produce both antigens codominantly and thus have antibodies to neither (and are tolerant to blood of any type). They can receive any blood, but their blood is toxic for all but the small fraction of receivers who are also AB (Figure 2.1).

COMPLICATIONS IN BLOOD TYPING

Currently, a rapid test, done by mixing specific antibodies with the blood and observing where the cells agglutinate, can determine what blood group a person has. Individuals with Type A blood will react with the antibodies produced against A-type RBCs, B with antibodies raised against B-RBCs. AB individuals have both antigens and will react with both antibodies. Type AB blood smear will show aggregation in both the A and B panels, while Type O patients will not show aggregation in either panel (since these RBCs lack both antigens).

Trauma physicians call for O^- blood, even though the blood type of the patient may already be known or quickly checked. This is because blood type agglutination assays may give unclear results, especially in an emergency situation (those with rare blood types should consider wearing an appropriate bracelet, as even in a hospital, it can take 15–30 minutes to type for ABO/Rh). Aggregation reactions are also sensitive to temperature[5] and pH, conditions that are difficult to control, for example after a car accident or on the battlefield.[6] Agglutination (sometimes called rosette formation, although the rosette shape is only seen under a microscope) may not be obvious in people who have weak expression of the A or B antigen due to point mutations, deletions or insertions in the ABO gene. One can sequence the ABO gene to determine the probable blood type, usually together with genes for other antigens that are expressed at lower concentration in cells. A single gene encodes the A or B galactosyltransferase enzymes (GATA, GATB) that add the A or B type determining final sugar group to the H antigen (Figure 2.2). Although GATA and GATB occur genetically as distinct enzymes, they differ from each other by only four amino acids.[7] Just these four differences determine whether a person has GATA, which binds preferentially to UDP-GalNAc (and transfers this to the H core to make the A-type antigen) or GATB, which will preferentially bind and transfer UDP-Gal (to make the B-type antigen). However, relating genetics to the observed phenotype may be difficult.

Sequencing of the ABO gene, located on chromosome 9, in different individuals revealed that mutations at the nucleic acid or protein level can greatly affect a person's blood type. Both forms of the galactosyltransferases may occur in AB individuals, who may inherit a gene from each parent. Alternatively, the AB_{weak} phenotype may also be due to mutations in the GAT protein allowing it to (weakly) accept and transfer both conjugated forms of UDP.[8] The mutations that allow this to happen also greatly weaken the enzyme, so the overall projection of antigen on the RBC surface is much lower, leading possibly to identifying their blood as O type. Also, blood types are not always stable. For example, there may be no or reduced expression of A or B antigens in those with leukemias or other myelodysplastic syndromes,[9] again making the blood type difficult to determine especially when random mutations occur.[10]

Comparing the sequence of the ABO gene in different populations has revealed many different sequences and versions of this basic blood system. If one went far back enough in time, the most probable blood type would be A, consistent with its high frequency in most populations today. B is a newer (and simpler) variant of A, occurring less frequently. While there is some disagreement about the order of occurrence, it is generally thought that the AB type is the most recent. Many (at least 60) different types of "null alleles" in the ABO gene have been found to lead to O type, where neither GATA or GATB is produced (or active). Several of these nucleotide changes are ancient, by some estimates predating the history of the human race by several million years.[11]

WHY ARE ABO INCOMPATIBILITY REACTIONS IMMEDIATELY SEVERE EVEN IN PEOPLE WHO HAVE NEVER RECEIVED BLOOD?

Normally, as described below for the Rh factor and other proteins (Figure 2.3 is an overview of some of the antigens and Hemoglobin [Hb] types expressed on RBCs discussed here), the body must be exposed before an antibody is produced. Indeed, directing immune recognition to "non-self" antigens of a pathogen is at the heart of vaccine design.[12] However, the original typing tests (and transfusion failures) depended on the fact that, even without exposure, those producing RBCs carrying the A antigen have antibodies in their sera against B antigens, and those producing neither (O type) have antibodies to both (see the Table in the bottom part

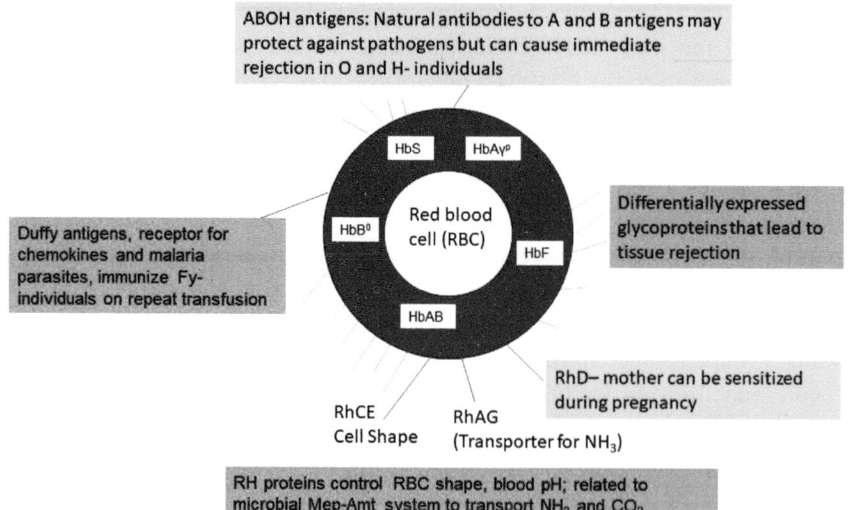

FIGURE 2.3 RBCs are coated with multiple copies of genetically encoded glycoprotein antigens. Both the type of hemoglobin expressed and variable antigens on the RBC surface affect immune recognition and even pathogen entry. Normal HbAB is a tetramer of 2 α and 2 β chains; HbF, produced in the fetus until shortly after birth, consists of 2 α and 2 γ chains. Hb mutants include HbS, sickle cell, with one amino acid mutation (Glu6 to Val in the β chain), Hbβ^0, beta thalassemia and HbAγ^p, where fetal hemoglobin persists (hydroxyurea treatment of sickle cell disease also upregulates this). Some individuals have HbS with Hbβ^0. The orange and blue lines indicate differentially expressed antigens (see https://ftp.ncbi.nlm.nih.gov/pub/mhc/rbc/Final%20Archive) that mediate immune reactivity. Yellow distinguishes the ABOH and Rh$^{+/-}$ blood types, which can lead to immediate or delayed immune reaction with mismatched blood. Orange indicates other potentially sensitizing antigens that can cause hemolytic reactions after exposure through pregnancy, repeated blood transfusions or tissue transplantation.

of Figure 2.1). These "natural antibodies" arise during the first months of life, stimulated by external antigens that resemble the A and B antigens[13] and are fully expressed by two to four years of age. Such antibodies may be important to protect against zoonotic (animal) viruses which make carbohydrate antigens according to their source.[14] Presumably anti-A, -B antibodies are suppressed (or tolerance to the A or B antigens is induced) during early development of the immune system in persons expressing these antigens. Suppression does not occur in an individual of Type O, so the natural antibodies persist. Putting A, B or AB blood (or tissue[15]) into an O-type person will lead to immediate and potentially lethal hemolytic immune reaction due to their anti-A and -B antibodies.

The ABO system is found in all humans, and similar systems are found in animals as well. Cats, for example, are about 70% A type. Yet there is no indication that O type has any deleterious effect on its carriers and may even protect them against cancer.[16] Further, their antibodies against A and B antigens may provide O-type individuals protection against some viruses, which may use the sugar chains as receptors to bind to cells. Indeed, several diseases, including some forms of viral infection and lung disease,[17] may be more aggressive in A-type people because their immune system may not recognize similar polysaccharide antigens expressed by the infecting virus. However, this difference due to blood type is so small it is only seen when thousands of patients are compared. If the mother does not provide such antibodies to her unborn child, this may also cause some problems in resistance. Another study found that babies who expressed AB antigens (where the mother probably lacked those natural antibodies) is linked to more neonatal problems.[18] This suggests that, if only the strongest survive, a novel pathogen neutralized by natural antibodies may select for blood type O (as alleles of the endoplasmic reticulum aminopeptidase 2 [ERAP2] gene seen to have been selected for, as described in Chapter 1). The GAL A or GAL B genes may be deleted in future generations, or so mutated that they will join the thousands of other pseudogenes that dot the human genome.

DEVELOPING ANTIBODIES TO VARIABLE ANTIGENS COMPLICATES REPEAT TRANSFUSIONS AND TISSUE TRANSPLANTATION

In addition to the ABO blood groups, which are determined by sugar group additions to membrane proteins, many other blood proteins that vary ("show multiple alleles") across individuals can induce "isoimmunization" (also called alloimmunization) in the receiver.[19] A list of known problematic proteins whose alleles must be matched for successful tissue transplantation fills a whole database.[15] Unlike the ABO (or antigens of the groups M, P, Lewis, and Li not discussed here), where antibodies arise naturally during development, a person typically becomes sensitized to these proteins after being exposed to foreign blood or tissue. Development of an immune response to these variable antigens leads to reactions during pregnancy, repeated blood transfusions or tissue rejection after transplantation. The reactions can occur within a few minutes or over several months. Modern blood typing automatically includes tests for the Rh factor and some other clinically significant RBC antigens. Even with careful matching to their donor for a series of known variable alleles, those receiving an organ or tissue transplant usually must take an immunosuppressant for the rest of their lives to prevent rejection of foreign tissue.

THE RH PROTEIN COMPLEX

The most well-known of the RBC expressed, conditionally toxic proteins, is the Rh factor, so named because it was first identified in Rhesus monkeys. Most people are RhD^+; about 15% of Americans are designated Rh^-, meaning they do not have this protein on their RBCs. Unlike the ABO proteins, where A antigen expression may have little or even a deleterious function, the Rh family of proteins has been implicated in transporting gases in and out of the red cell membrane (more on that later). Yet RhD^- people do not have noticeable health effects, at least as far as their RBC function and life span. The problem is first seen when they come in contact with RhD^+ blood, most commonly during pregnancy.

Unlike the rapid reactions to the ABO factors, due to the preexisting antibodies, developing a strong antibody response to Rh antigen after exposure of an Rh^- person to Rh^+ blood is a longer term process as the antibodies are induced, not innate. Thus, an Rh^- woman (about 15% of the population) may have an Rh^+ child with an Rh^+ man (about 85% of the population). Her first pregnancy may be normal, assuming there is no bleeding throughout. If exposure to an Rh^+ baby's blood during birth sensitizes the mother, her second pregnancy may end prematurely as the anti-RhD protein antibodies formed after the first pregnancy invade the placenta and attack the RBCs of an Rh^+ fetus. This leads to fetal anemia and in the worst case, hemolytic disease and possibly miscarriage. The only treatment for incompatibility, if the mother has become sensitized, is blood transfusion to the baby, who may be born jaundiced (due to the breakdown products of the extra blood the little body has trouble clearing).

To prevent sensitization, antibodies to the Rh factor are given to the mother in injections after birth. These bind to and neutralize RhD and prevent the protein from inducing the woman to make her own antibodies to it, thus making her tolerant to the RhD in the next pregnancy. Women who are Rh^- are given the antibodies at 28 weeks of pregnancy, after any episode of vaginal bleeding and within 72 hours after delivery of a baby who has Rh^+ blood, even after a miscarriage or an abortion. The treatment reduces the risk that fetal RBCs will be destroyed in subsequent pregnancies from about 12–13% (without treatment) to 1–2%.

Throughout this discussion, only the RhD protein was mentioned. RhD is one part of an Rh group of related proteins. While the galactosyltransferases that control the A and B antigen addition differ only slightly, the components of the Rh complex vary in their amino acid sequences[20] and are encoded by different genes. As with the ABO system, the Rh proteins are expressed on many other tissues throughout the body and need to be matched during tissue transplantation typing. To reach the cell surface, RhD must form a complex with a related protein, RhAG.

Most Rh^- people have complete deletion of the gene. There is also "partial D" phenotype, where the individual lacks part of the protein chain required, resulting in the ability of the individual to make anti-RhD when exposed to RhD antigen. Other mutations can lead to low expression of the protein, making careful blood typing essential, as otherwise "weak" RhD^+ blood might be given to an Rh^- person. While this will not induce an immediate response in the short term, if the person has never been exposed to RhD^+ cells before, this may lead to future problems. Other mutations may have much more serious consequences. For example, some have an "Rhneg" phenotype (distinct from RhD^-), meaning they do not express any part of the Rh family. Rhneg individuals are typically anemic, and they can only obtain blood from other Rhneg people. Their anemia is due to their RBCs being misshapen and easily lysed, a fact attributed to the lack of expression of the small RhCE protein. They may also have trouble with acidification of blood, as discussed below.

WHAT GOOD IS THE RH PROTEIN?

RhD^- people seem to have no negative effects other than the possibility of immune sensitization described above.[21] Thus, one might assume that the Rh proteins, like the AB antigens, are an evolutionary vestige within the genome, whose function was lost or turned to a purely cell structure maintaining one during the eons of human development. This theory was contradicted when sequencing of similar genes revealed that Rh proteins have ancient ancestors going back to the earliest life forms on earth. This revealed the Rh complex's basic and necessary functions in RBCs. The human Rh proteins are related in sequence and structure to the Mep-Amt family of proteins in yeast and bacteria that control the export and import of ammonia (NH_3).[22,23] Indeed, the Rh protein complex from either RBCs or the kidney[24] could help transfer ammonia across mammalian and bacterial membranes,[25] suggesting it has a role in regulating blood pH. The related smaller protein RhCE has a role in maintaining the shape of RBCs. The RhAG protein, essential for the expression of the Rh proteins on RBCs, has retained its original function over billions of years of evolution. When the gene for the human RhAG protein was expressed in yeast, it transported toxic ammonia out of the cell into the medium. Further, known human mutations of RhAG, associated with problems in controlling blood pH and ammonia toxicity, prevented this transport.[23] This confirmed the essential role of Rh protein complexes in removing problematic gases produced during cellular metabolism.

There was yet another explanation for what the Rh complex was specifically doing on blood cells: it has evolved to facilitate the transport of another, conditionally toxic gas: carbon dioxide (CO_2).[26] Carbon dioxide at low concentrations is innocuous, the gas that makes bubbles in club soda and other soft drinks. But at high concentrations, it can become a lethal toxin. RBCs must transfer oxygen (O_2) to all tissues and, as importantly, quickly move the metabolically generated CO_2 to the lungs, where it can leave the body. Hemoglobin, the major protein in RBCs, changes structure with environment to allow it to carry either O_2 or CO_2. The Rh protein can help bring the gas to Hb in this process, by helping CO_2 move through membranes.

MALARIA SELECTION FOR HB TYPE AND RBC ANTIGENS

Significant matching of other variably expressed blood cell proteins is especially important for those with sickle cell anemia (here referred to as sickle cell disease, or SCD), as well as others who require frequent transfusions.[27] In SCD, a single amino acid change in the β chain of hemoglobin (resulting in a protein called HbS) causes the protein to aggregate under certain conditions. This causes the RBCs to deform to a characteristic sickle shape, eventually resulting in the cells lysing.[28] In addition to the reduced ability of blood to carry oxygen, cell detritus can accumulate to cause painful joints and block blood vessels, causing potentially catastrophic organ infarcts and strokes. About 100,000 Americans have SCD, predominantly those with African ancestry. One study found that mutations at and around this locus in the human Hb gene occur more frequently in those of African descent, further illustrating the importance of this conditionally toxic protein for reasons other than blood function.[29] SCD can be inherited even from non-anemic parents with "sickle cell trait" who are heterozygous (HbA/S), with one gene copy that contains the HbS mutation. In parts of Africa where malaria is a debilitating and/or fatal infection, there are about 6 million HbS people with at least one copy of the HbS gene.

The phenotype probably became fixed in regions where malaria is common and children are often undernourished. Under these conditions, a child with the HbS trait might survive long enough to have offspring, while his siblings with two "normal" genes for HbAB* succumbed to a combination of malnourishment and malaria. The child's RBCs would resist infection

* The AB here refers to α and β chains of the Hb tetramer, not to the A-type antigen on surface of the RBCs in Figures 2.1 and 2.2. Another type of Hb is HbF, which has two γ subunits in place of the β ones in adult Hb. HbF has very high oxygen transport capability and is produced in the fetus. HbF may persist after birth in individuals with HbS or other Hb mutants; it can also be induced by the hydroxyurea treatment used for sickle cell disease.

by the malaria parasite, due to the aggregation of hemoglobin in the cell. Further, malnourishment leads to anemias and a lower concentration of RBCs in the blood stream, meaning that cells are less likely to clump and cause the other debilitating effects of SCD.

Today, transfusions are used to treat SCD crises, with the goal of rapidly increasing the oxygen-transporting ability of blood by introducing RBCs containing normal Hb and decreasing the concentration of HbS-RBCs to 30% or lower. However, over time these transfusions can lead to other complications, such as too much copper or other trace metals in the blood, and serious clinically significant antibody formation to RBC antigens. The "Duffy" antigens (named for a patient in whom antibodies to these proteins were first identified) are often not expressed on RBCs in populations from malaria-prone regions, such as Madagascar[30] and many areas of Africa, while present in most Europeans and Asians.[31] The proteins (called FyA or B) are binding sites for *Plasmodium falciparum* (a malaria parasite), enabling the infection of RBCs. The Duffy antigens are glycoproteins that can serve as chemokine binders, although they lack the G-protein coupling (which will be discussed in Chapter 5) that mediate cytokine[32] signaling after binding to their receptors. Removing inflammatory chemokines such as IL-8 from the blood stream may help control an overactive immune response (see Chapter 6), but Duffy^{+} phenotype has not been shown to offer Covid-19 patients any protection.[33]

As blood typing does not typically note the race of the donor, giving Duffy^{+} blood (i.e., that most commonly present in European and American blood banks) to a black SCD patient (most of whom are Duffy^{-} [34]) can lead to immunization against these antigens. *It is thus essential for people from all ethnicities to contribute to their local blood banks, to have sufficient supplies for all types of emergencies.*

In addition to sickle cell syndromes, other conditions, such as thalassemia (lacking the B chain of hemoglobin [HbB^{0}])[35] and after sensitization to the Rh factor require repeated blood transfusions and accurate matching of many different blood proteins. Unfortunately, the database describing the many known mutations that can affect immunogenicity of blood proteins,[36] dbRBC, is not presently maintained. It is now only available as individual files: https://ftp.ncbi.nlm.nih.gov/pub/mhc/rbc/Final%20Archive/.

CONTROLLING THE IMMUNE SYSTEM IN ORGAN TRANSPLANTATION

Organ transplantation is dependent on our modern ability to modulate the immune system. Many antigens, including the ABOH and Rh systems, are expressed in many different cells in the body. As noted above, treatment to control the immune response to the Rh factor is primarily palliative (giving the fetus blood transfusions to overcome the antibody attack from the maternal antibodies) or preventative (administering antibodies to the Rh factor, to mask it from the immune system). In organ transplantation, the major goal of the therapy is to prevent the formation of antibodies to the foreign tissue in the first place, as once they are induced, the graft will be rejected. Cyclophilin, one of the first agents to be used to prevent tissue rejection, might be considered a conditional toxin, as it blocks an essential section of the immune system, namely the activation of T-cells to induce B-cells to make antibodies. Those who are being treated are immunosuppressed, unable also to make antibodies to pathogens. However, they also will not make antibodies to any unmatched antigens on the foreign tissues, preventing immune attack on the graft that causes rejection.

While cyclophilin may still be used after transplant therapy to block the formation of antibodies, a small molecule treatment called Tacrolimus (FK506) is more generally prescribed, as it is easier to make and more potent (i.e., has a higher specific activity). There are a variety of other compounds used as immunosuppressors with different modes of action (see https://www.aaaai.org/Conditions-Treatments/Related-Conditions/immunosuppressive for a list). Some of these will be discussed in later chapters.

The target of both cyclophilin and FK506 is a protein called calcineurin. Binding to this protein prevents phosphorylation of a protein that leads to T-cell activation. This stops them from making cytokines, including a master controller, another conditionally toxic protein, interleukin 2 (**IL-2**). This cytokine not only stimulates T-cell development but also plays a role in achieving tolerance, as will be discussed in Chapter 6.

CONCLUSIONS

Giving blood, the essential carrier of oxygen to all tissues and remover of metabolic toxins, can save a life, or if mismatched, can kill. The story of ABOH, Rh and the many other antigens expressed on the surface of RBCs and tissues illustrates how a normally protective immune response can instead induce destruction. Malaria, other infectious diseases, and malnourishment have played a role in shaping the antigens expressed on cells and even the Hb type within RBCs. Success in blood transfusions and tissue transplantation requires preventing good and useful proteins, conserved since the dawn of existence, from becoming killers. The best way to prevent transfusing “the wrong blood” in an emergency is having blood type information easily available. Those with very rare blood types should consider wearing a special alert bracelet and people from all ethnicities should be encouraged to donate blood to local banks.

As discussed in the next chapters, allergenic proteins and even our daily bread provide further examples of how the immune system can perceive perfectly good proteins as toxic threats.

For more details, please see the following references.

REFERENCES

1. Mourant AE. The discovery of the anti-globulin test. Vox Sang. 1983; 45(2):180–3. doi: 10.1111/j.1423-0410.1983.tb01902.x. PubMed PMID: 6351437.
2. Wagner GK, Pesnot T, Palcic MM, Jorgensen R. Novel UDP-GalNAc derivative structures provide insight into the donor specificity of human blood group glycosyltransferase. J Biol Chem. 2015;290(52):31162–72. Epub 20151102. doi: 10.1074/jbc.M115.681262. PubMed PMID: 26527682; PMCID: PMC4692239.
3. Hosoi E, Hirose M, Hamano S. Expression levels of H-type alpha(1,2)-fucosyltransferase gene and histo-blood group ABO gene corresponding to hematopoietic cell differentiation. Transfusion. 2003;43(1):65–71. doi: 10.1046/j.1537-2995.2003.00277.x. PubMed PMID: 12519432.
4. Shaik L, Ravalani A, Devara J, Rathore SS, Singh R. Secondary postpartum hemorrhage presenting with Bombay blood group: a case report. Cureus. 2020;12(8):e9758. Epub 20200815. doi: 10.7759/cureus.9758. PubMed PMID: 32944472; PMCID: PMC7489786.
5. Schein CH. Solubility as a function of protein structure and solvent components. Biotechnology (N Y). 1990;8(4):308–17. doi: 10.1038/nbt0490-308. PubMed PMID: 1369261.
6. Levine P. Historical perspectives before 1945 – updated. Prog Clin Biol Res. 1980;43:3–11. PubMed PMID: 6999506.
7. Yamamoto F, Hakomori S. Sugar-nucleotide donor specificity of histo-blood group A and B transferases is based on amino acid substitutions. J Biol Chem. 1990;265(31):19257–62. PubMed PMID: 2121736.
8. Marcus SL, Polakowski R, Seto NO, Leinala E, Borisova S, Blancher A, Roubinet F, Evans SV, Palcic MM. A single point mutation reverses the donor specificity of human blood group B-synthesizing galactosyltransferase. J Biol Chem. 2003;278(14):12403–5. Epub 20030115. doi: 10.1074/jbc.M212002200. PubMed PMID: 12529355.
9. Zhang W, Liu J, Zhang W, Zhuang Y. The potential association of the transcription levels of the ABO gene with the disease phases in AML patients. Transfus Apher Sci. 2017;56(5):719–22. Epub 20170901. doi: 10.1016/j.transci.2017.08.004. PubMed PMID: 28939368.
10. Hayakawa A, Sano R, Takahashi Y, Kubo R, Harada M, Omata M, Yokohama A, Handa H, Tsukada J, Takeshita H, Tsuneyama H, Ogasawara K, Kominato Y. RUNX1 mutation in a patient with myelodysplastic syndrome and decreased erythrocyte expression of blood group A antigen. Transfusion. 2020;60(1):184–96. Epub 20191216. doi: 10.1111/trf.15628. PubMed PMID: 31840280.

11. Roubinet F, Despiau S, Calafell F, Jin F, Bertranpetit J, Saitou N, Blancher A. Evolution of the O alleles of the human ABO blood group gene. Transfusion. 2004;44(5):707–15. doi: 10.1111/j.1537-2995.2004.03346.x. PubMed PMID: 15104652.
12. Baker WS, Negi S, Braun W, Schein CH. Producing physicochemical property consensus alphavirus protein antigens for broad spectrum vaccine design. Antiviral Res. 2020;182:104905. Epub 20200812. doi: 10.1016/j.antiviral.2020.104905. PubMed PMID: 32800880.
13. Arend P. "Natural" antibodies and histo-blood groups in biological development with respect to histo-blood group A. A perspective review. Immunobiology. 2011;216(12):1318–21. Epub 20110527. doi: 10.1016/j.imbio.2011.04.008. PubMed PMID: 21798618.
14. Galili U. Human natural antibodies to mammalian carbohydrate antigens as unsung heroes protecting against past, present, and future viral infections. Antibodies (Basel). 2020;9(2). Epub 20200622. doi: 10.3390/antib9020025. PubMed PMID: 32580274; PMCID: PMC7344964.
15. Zhang Q, Rajalingam, R., Cecka, J.M. Reed, E.F. Chapter 89 – ABO, tissue typing, and crossmatch incompatibility. Transplantation of the Liver (Third Edition): Elsevier; 2015.
16. Arend P. ABO phenotype-protected reproduction based on human specific alpha1,2 L-fucosylation as explained by the Bombay type formation. Immunobiology. 2018;223(11):684–93. Epub 20180721. doi: 10.1016/j.imbio.2018.07.015. PubMed PMID: 30075871.
17. Reilly JP, Meyer NJ, Shashaty MG, Anderson BJ, Ittner C, Dunn TG, Lim B, Forker C, Bonk MP, Kotloff E, Feng R, Cantu E, Mangalmurti NS, Calfee CS, Matthay MA, Mikacenic C, Walley KR, Russell J, Christiani DC, Wurfel MM, Lanken PN, Reilly MP, Christie JD. The ABO histo-blood group, endothelial activation, and acute respiratory distress syndrome risk in critical illness. J Clin Invest. 2021;131(1). doi: 10.1172/JCI139700. PubMed PMID: 32931480; PMCID: PMC7773362.
18. McMahon KE, Habeeb O, Bautista GM, Levin S, DeChristopher PJ, Glynn LA, Jeske W, Muraskas JK. The association between AB blood group and neonatal disease. J Neonatal Perinatal Med. 2019;12(1):81–6. doi: 10.3233/NPM-17115. PubMed PMID: 30347622.
19. Wafford TR, Walker LP. Prevalence of Rh, Kell, Kidd, Duffy, and MNS antigens in the Hispanic donor population of South Texas. Immunohematology. 2022;38(2):43–50. Epub 20220705. doi: 10.21307/immunohematology-2022-040. PubMed PMID: 35852064.
20. Lupo D, Li XD, Durand A, Tomizaki T, Cherif-Zahar B, Matassi G, Merrick M, Winkler FK. The 1.3-A resolution structure of *Nitrosomonas europaea* Rh50 and mechanistic implications for NH3 transport by Rhesus family proteins. Proc Natl Acad Sci USA. 2007;104(49):19303–8. Epub 20071121. doi: 10.1073/pnas.0706563104. PubMed PMID: 18032606; PMCID: PMC2148285.
21. Dahlen T, Clements M, Zhao J, Olsson ML, Edgren G. An agnostic study of associations between ABO and RhD blood group and phenome-wide disease risk. Elife. 2021;10. Epub 20210427. doi: 10.7554/eLife.65658. PubMed PMID: 33902814; PMCID: PMC8143790.

22. Huang CH, Peng J. Evolutionary conservation and diversification of Rh family genes and proteins. Proc Natl Acad Sci USA. 2005;102(43):15512–7. Epub 20051014. doi: 10.1073/pnas.0507886102. PubMed PMID: 16227429; PMCID: PMC1266151.
23. Deschuyteneer A, Boeckstaens M, De Mees C, Van Vooren P, Wintjens R, Marini AM. SNPs altering ammonium transport activity of human Rhesus factors characterized by a yeast-based functional assay. PLoS One. 2013;8(8):e71092. Epub 20130813. doi: 10.1371/journal.pone.0071092. PubMed PMID: 23967154; PMCID: PMC3742762.
24. Ripoche P, Bertrand O, Gane P, Birkenmeier C, Colin Y, Cartron JP. Human Rhesus-associated glycoprotein mediates facilitated transport of NH(3) into red blood cells. Proc Natl Acad Sci USA. 2004;101(49):17222–7. Epub 20041130. doi: 10.1073/pnas.0403704101. PubMed PMID: 15572441; PMCID: PMC535366.
25. Westhoff CM. The structure and function of the Rh antigen complex. Semin Hematol. 2007;44(1):42–50. doi: 10.1053/j.seminhematol.2006.09.010. PubMed PMID: 17198846; PMCID: PMC1831834.
26. Huang CH, Ye M. The Rh protein family: gene evolution, membrane biology, and disease association. Cell Mol Life Sci. 2010;67(8):1203–18. Epub 20091202. doi: 10.1007/s00018-009-0217-x. PubMed PMID: 19953292.
27. Wilkinson K, Harris S, Gaur P, Haile A, Armour R, Teramura G, Delaney M. Molecular blood typing augments serologic testing and allows for enhanced matching of red blood cells for transfusion in patients with sickle cell disease. Transfusion. 2012;52(2):381–8. Epub 20110809. doi: 10.1111/j.1537-2995.2011.03288.x. PubMed PMID: 21827505.
28. Mangla A, Ehsan M, Agarwal N, Maruvada S, Doerr C. Sickle Cell Anemia (Nursing). StatPearls: Treasure Island, FL, 2022.
29. Melamed D, Nov Y, Malik A, Yakass MB, Bolotin E, Shemer R, Hiadzi EK, Skorecki KL, Livnat A. De novo mutation rates at the single-mutation resolution in a human HBB gene region associated with adaptation and genetic disease. Genome Res. 2022;32(3):488–98. Epub 20220114. doi: 10.1101/gr.276103.121. PubMed PMID: 35031571; PMCID: PMC8896469.
30. Hodgson JA, Pickrell JK, Pearson LN, Quillen EE, Prista A, Rocha J, Soodyall H, Shriver MD, Perry GH. Natural selection for the Duffy-null allele in the recently admixed people of Madagascar. Proc Biol Sci. 2014;281(1789):20140930. doi: 10.1098/rspb.2014.0930. PubMed PMID: 24990677; PMCID: PMC4100512.
31. Howes RE, Patil AP, Piel FB, Nyangiri OA, Kabaria CW, Gething PW, Zimmerman PA, Barnadas C, Beall CM, Gebremedhin A, Menard D, Williams TN, Weatherall DJ, Hay SI. The global distribution of the Duffy blood group. Nat Commun. 2011;2:266. doi: 10.1038/ncomms1265. PubMed PMID: 21468018; PMCID: PMC3074097.
32. Schein CH. From interleukin families to glycans: relating cytokine structure to function. Curr Pharm Des. 2004;10(31):3853–5. doi: 10.2174/1381612043382512. PubMed PMID: 16381102.

33. Andreakos E, Abel L, Vinh DC, Kaja E, Drolet BA, Zhang Q, O'Farrelly C, Novelli G, Rodriguez-Gallego C, Haerynck F, Prando C, Pujol A, Effort CHG, Su HC, Casanova JL, Spaan AN. A global effort to dissect the human genetic basis of resistance to SARS-CoV-2 infection. Nat Immunol. 2022;23(2):159–64. Epub 20211018. doi: 10.1038/s41590-021-01030-z. PubMed PMID: 34667308; PMCID: PMC8524403.
34. Drasar ER, Menzel S, Fulford T, Thein SL. The effect of Duffy antigen receptor for chemokines on severity in sickle cell disease. Haematologica. 2013;98(8):e87–9. Epub 20130610. doi: 10.3324/haematol.2013.089243. PubMed PMID: 23753024; PMCID: PMC3729889.
35. Sarihi R, Amirizadeh N, Oodi A, Azarkeivan A. Distribution of red blood cell alloantibodies among transfusion-dependent beta-thalassemia patients in different population of Iran: effect of ethnicity. Hemoglobin. 2020;44(1):31–6. doi: 10.1080/03630269.2019.1709205. PubMed PMID: 32400249.
36. Patnaik SK, Helmberg W, Blumenfeld OO. BGMUT database of allelic variants of genes encoding human blood group antigens. Transfus Med Hemother. 2014;41(5):346–51. Epub 20140915. doi: 10.1159/000366108. PubMed PMID: 25538536; PMCID: PMC4264482.

Introduction to Chapters 3 and 4
Intolerance and Elimination Diets

> … since the disease appears in many instances to be a form of chronic neurosis, remedies which improve the stability of the nervous system may be employed- such as arsenic, phosphorous, and strychnia.
>
> **William Osler, Autumnal Catarrh (Hay fever) in *Principles and practices of medicine* (1st edition, 1892 by D. Appleton & Company, p. 478)**

Allergies and food intolerances have long been the subject of controversy: are they real or is it all in the patient's head? Lacking our current understanding of immunology and tools to measure antibodies in sera, doctors in the first part of the 20th century chose to assume the latter, prescribing nervous system "stabilizers" such as those recommended above (in a quote from one of the first widely used general medical textbooks in the United States). Today, doctors understand that food intolerances are truly immune-mediated diseases. Intolerance to wheat proteins can be documented by antibodies in the sera of celiac disease (CeD) sufferers, a diagnosis that can be further confirmed by endoscopy. Serum IgE levels to allergens, coupled with skin prick or oral challenge, can be used to confirm allergy.

These tests show clearly that foods consumed by most of the population, containing glutens (discussed in Chapter 3) or allergens such as peanuts or shrimp (Chapter 4), can be toxic to sensitized individuals. Chapter 3, "Minding the Ps and Qs of gluten", describes how the amino acid content distinguishes conditionally toxic wheat glutenins. Most people eat gluten, in products ranging from bread to sausages, with no ill effects. However, for individuals with celiac disease, eating gluten causes symptoms ranging from headaches to skin rash to chronic diarrhea and enhances their risk of colon cancer. Understanding the immune response to peptides containing high concentrations of the amino acids proline (P) and glutamine (Q) has enabled assays that can aid in diagnosing celiac disease, even without a biopsy to show intestinal damage.

DOI: 10.1201/9781003333319-3

While glutens have peculiar sequences that clearly distinguish them from other proteins, allergens, the topic of Chapter 4, are more difficult to categorize. Identifying similar proteins to known allergenic ones may help to define elimination diets for those with known sensitivities. Peanuts, tree nuts, shrimp, eggs and milk are common foods that can induce severe reactions in sensitized individuals. Today allergies are recognized as serious conditions. Early medical views were tainted by misogyny and questionable assertions by those recognized as experts. After much controversy in the early 20th century about hay fever being neurotic or "hysterical", the word allergy was coined. Public reports of deaths attributable to anaphylaxis from certain foods in the 1960s spurred efforts to develop treatments and define the molecular properties of allergenic proteins in the 1970s.[1] Understanding commonalities in the sequence and structure of proteins that induce anaphylactic responses provide clues to why some people are sensitized, and what areas of these proteins could be rendered less problematic by changing their sequences or structures.[2]

While we can now diagnose these syndromes immunologically, there are still no simple treatments. Once diagnosed, abstaining from problematic foods may be the only treatment available for sensitized people (Figure I.1).

FIGURE I.1 Elimination diets can be difficult. Those with sensitivities must avoid food (containing conditionally toxic proteins) that most people eat without problems.

There are many obstacles to this solution in a world where allergies or gluten sensitivity is only experienced by a fraction of the population. For individuals with celiac disease, the recommended treatment is to completely avoid eating gluten proteins found in wheat and other grains. Due to the omnipresence of gluten proteins in foods, celiac patients are often hospitalized to make sure that they really are eating gluten free. Similarly, the only treatment available for many allergy sufferers is to avoid all sources of the foods or pollen they are sensitized to. Better solutions will require an enhanced understanding of the characteristics of these conditionally toxic proteins.

MAINTAINING AN ELIMINATION DIET IS HARD AND OFTEN LONELY WORK

The major difficulty in following an elimination diet is sabotage by oneself – or other people. As noted in the quote above from Dr. Osler's early and well-respected medical textbook, allergies and food sensitivities were initially classified as neuroses. This attitude persists in many parts of the medical profession and general population. Part of the problem for those with sensitized immune systems is that the most dangerous proteins are only conditionally toxic.

While schools can make rules to protect individuals, they are often broken. A parent, whose children thrive on peanut butter and jelly on white bread sandwiches with a glass of milk, may find it hard to sympathize with the one classmate who can go into shock just breathing peanut fumes. Conversely, avoiding milk or wheat is tough for the lone child in the class who cannot eat the beautiful cupcakes Alice's mom sends in for her birthday. Children get hungry!

Even dining with beloved Aunt Eva can become a challenge, when she brings out a plate of fried chicken and does not understand the need to cut away her perfect, delicious crust. One could retort that Aunt Eva might spend more time exercising and less time massaging meat with buttermilk and flour before dropping it into hot oil. For Aunt Eva could be among the 50–70% of the US population who are heavier than they should be. She may have had her own problems maintaining an elimination diet. She may have tried to follow Atkins, for example, and wound up even heavier than before when she ate a cookie in addition to a large steak. If she and those around her would recognize the support needed by those on elimination diets, perhaps everyone, including those burdened with food allergies or overweight, would experience less struggle and more success.

The majority of the human race has trouble sticking to a diet. Yet somehow we assume that 3–8% of our population with food sensitivities should be able to control their diet, without substantial help and encouragement. Indeed, those requesting a special diet frequently meet with impatience, intolerance and covert sabotage.

In a famous study illustrating the difficulties of dietary self-control, children were tempted with a marshmallow, which they could eat immediately, or hold off for a time and get two marshmallows. Later, the investigators were interested in seeing whether the patient children who got the bigger treat were more than just hungry. They concluded that delay time in the test correlated with self-control indices. What is never really covered is that at least half of the children did not bother with the study at all (they took the marshmallow and went to play), ending the whole test early.

We all know that following a diet that allows one to eat what everyone else does, just less of it, is often beyond the capabilities of most people. Yet those with food sensitivities must eliminate all foods containing wheat, for example, or any form of nuts. The general population does not understand, or particularly care about, food sensitivities. They may not believe they are real or dangerous. People following an elimination diet for medical reasons thus fear being lied to by people offering or selling food. A chef in a restaurant has a set way and time management for preparing their entrees. All their gravies and gumbos were probably simmering since the early morning hours, long before the patient with celiac disease shows up requesting a gluten-free plate. The waiter who brings the order has little time to dive into the question of whether the sauce was made with cornstarch or wheat flour. A busy mother probably cannot remember if she used a margarine that contains milk when she prepared rice crispy snacks. The situation is worse in fast-food establishments (and most school cafeterias) which rely on pre-prepared items.

So those who must avoid gluten or allergens are often doomed to a lonely life, doing their own cooking, eating at home, bringing their own lunches. Parties become a challenge, when eating a single canape or piece of cake might lead to gastrointestinal agony later, or even anaphylactic shock in the middle of the festivities.

Even the most astute patient faces an insidious problem: one gets hungry. Even when one is not hungry, the urge to eat a tasty morsel is probably built into our DNA. This problem is shared, by Aunt Eva and all those others who know very well what they should not eat but eat it anyway, and consequences be damned. Unfortunately, for those extremely allergic to peanuts or shrimp, for example, those consequences can be deadly.

MODERN PROTEIN SOURCES MAY BE DESIGNED TO AVOID ALLERGENS OR TOXINS

Many now understand that mankind's survival depends on the availability of high-quality proteins to feed a continuously growing world population. Immune assays clearly indicate foods that are toxic to some individuals. However, the popular press has written extensively about both celiac disease and anaphylactic reactions to certain foods, leading to a view by many that wheat and nut products should be avoided. Thus, in addition to the fraction of the population who suffer from celiac disease and food allergies, many individuals now shun all gluten, nut and soy protein containing products. A person wanting to eat vegan will have difficulty putting together a diet with sufficient protein (30–60 g/day) that contains none of those proteins, in addition to no meat or dairy!

Fortunately, modern science (and genetic engineering) may offer a solution. It is now possible, using food-grade ingredients, to grow large quantities of delectable fungi (think mushrooms, not slime molds) in solution that could be a useful meat alternative, low in saturated fat and high in essential amino acids. There is now a growing and potentially very profitable market for these products as they steadily improve in taste and variety. Most importantly, these engineered proteins can be made hypoallergenic, easily digestible, and free of "FODMAPS" or gluten. Modern artificial meats start with vegetable sources that are already suitable for human consumption. Although these may still cause problems for those with soy, pea or wheat allergies, if we understand the properties of the proteins that contribute to allergenicity, we should be able to encode hypoallergenic versions. There is a continuous search for non-gluten, hypoallergenic proteins that can be used to construct more textured products. In addition to making certifiably gluten and allergen-free foods, delectable protein sources can be produced for those with diseases such as phenylketonuria (the inability to completely remove the degradation products of phenylalanine from the body, leading to toxic buildup) or to obtain an umami flavor in the absence of monosodium glutamate (MSG).

CONCLUSIONS

Chapters 3 and 4 will describe how different conditionally toxic proteins in food and pollen trigger immune system reactions in sensitized individuals. Although elimination diets are very hard to maintain, the major "treatment" for celiac disease and allergies is currently avoidance of specific triggering sources. Eliminating foods such as wheat and nuts requires very careful food preparation, as these are omnipresent in the food chain and consumed with pleasure by most of the population. New methods may allow one to produce foods that are safe for all to eat and provide cheap proteins for a hungry world. But in the meantime, everyone should understand that some people have true food sensitivities, not neuroses.

REFERENCES

1. Schein CH, Negi SS, Braun W. Still SDAPing along: 20 years of the structural database of allergenic proteins. Front Allergy. 2022;3:863172. Epub 20220322. doi: 10.3389/falgy.2022.863172. PubMed PMID: 35386653; PMCID: PMC8974667.
2. Schein CH, Ivanciuc O, Midoro-Horiuti T, Goldblum RM, Braun W. An allergen portrait gallery: representative structures and an overview of IgE binding surfaces. Bioinform Biol Insights. 2010;4:113–25. Epub 20101011. doi: 10.4137/BBI.S5737. PubMed PMID: 20981266; PMCID: PMC2964044.

3 Minding the Ps and Qs of Gluten

OVERVIEW

1. Wheat and barley flours are cheap to produce, shelf stable and serve many different roles in food texture and taste. This means they are everywhere in the food chain.
2. Gluten proteins found in grains (wheat, rye and barley) are conditionally toxic proteins to sensitive individuals.
3. Glutens such as wheat gliadin contain long stretches of the amino acids glutamine (Q) and proline (P).
4. This composition, different from most animal proteins, means they lack essential amino acids.
5. Gluten proteins are partially insoluble and processing (kneading and baking) makes them more so.
6. The long stretches of P and Q residues in their amino acid sequence are incompletely digested by proteases in stomach acid or in the duodenum, leaving long peptides that are prone to aggregation.
7. Transglutaminase enzymes can change the Q residues of the peptides to glutamate, leading to peptides that can trigger a T-cell-based immune response.
8. Long polyQ stretches are also found in human proteins implicated in neurological syndromes, Huntington disease (HD) and ataxias.
9. The primary current treatment for CeD is to avoid eating gluten, despite the problems in maintaining an elimination diet.

It is estimated that 1% of the American population does not digest the gluten proteins in wheat and barley very well. As most of the population enjoys eating bread, pasta, pastries and wheat thickened sauces, someone announcing they are "gluten intolerant" will not be met with universal sympathy and acceptance, as described in the previous section of this book. Yet those with celiac disease (CeD) can experience severe intestinal pain if they give in and eat "just a little bit" of wheat containing foods.

DOI: 10.1201/9781003333319-4

A classic CeD patient is young and may present with some combination of symptoms, such as weight loss, diarrhea, vomiting, perhaps a skin rash, infertility or amenorrhea and frequent headaches. Their blood work may indicate vitamin or iron deficiency or even anemia. But other CeD patients may be over 60 and may complain of constipation, headache, pains in their head and feet, and in many cases, simply trouble concentrating or thinking clearly. One thing all CeD patients have in common: their symptoms are alleviated when they eliminate gluten.

They often meet their diagnosis with relief, finally knowing what is causing their various symptoms. However, to control CeD, they are condemned to drink their coffees without a croissant or donut on the side, eat risotto when they really want a big bowl of spaghetti and question every waiter about any dish with a sauce. Even then, they must live with the physical effects when the chef used wheat flour in his sauce crème without notifying the wait staff.

Meanwhile, they are surrounded by wonderful dishes that incorporate wheat, enjoyed by most of the world's population. The smell of fresh bread perfumes the streets around French and German bakeries. Asians slurp ramen. Greek and Middle Easterners eat pita. Moroccans use m'semen (semolina flatbread) like a fork. Indians wipe down their plates with naan.

Those who are gluten intolerant must scrutinize every detail of food preparation, knowing that every time they eat a classic gravy or gumbo, traditionally prepared with a flour and fat base, they risk recurring symptoms. No wonder one of the psychiatric symptoms of CeD is anxiety! They have the choice of revealing their CeD when their bosses order pizza or sub sandwiches for a lunch meeting or being branded by their coworkers as being "picky about food". Many CeD patients (and those with allergies, the topic of the next chapter) avoid eating in all restaurants and restrict their dining out to the homes of all but the most trusted of friends.

The ultimate diagnosis for CeD, the most severe wheat-related syndrome, is a biopsy of the upper digestive tract while the patient is continuing to eat wheat. However, the doctor may try a less invasive way to diagnose the disease, by measuring specific antibodies in the blood (see below) or simply telling the patient to avoid gluten-containing foods for a while to see if their symptoms improve. Going "gluten free" is easier than it was decades ago, but as the examples above illustrate, it is neither easy nor inexpensive.

GLUTENS HAVE DISTINCTIVE PROPERTIES

Gluten is primarily a mixture of two types of proteins called gliadins and glutenins. It is in the sticky mass that is hard to remove from fingers when kneading bread dough (rubbing hands with some olive oil before starting may avoid this). To isolate gluten, one can take a ball of pizza dough and wash it many times with water to remove the starch and water-soluble proteins. Glutens are in the sticky mass that is left. If that wet sticky ball is extracted with 120 proof moonshine or some other alcohol, one obtains a cloudy liquid containing gliadins, and an insoluble mass consisting primarily of glutenin.[1] The gliadins are also called prolamins and are related to some of the allergenic proteins discussed in the next chapter.

GLUTEN PROTEINS HAVE UNBALANCED SEQUENCES LACKING ESSENTIAL AMINO ACIDS

All proteins contain some mixture of 20 different amino acids, each of which can be represented by a single letter (Table 3.1). Gluten is an outlier, as the proteins that make it up contain disproportionate amounts of two amino acids, proline (P) and glutamine (Q), in characteristic repeats. Highlighting these two amino acids in one of the gliadins (Figure 3.1):

MKTLLILTILAMAITIGTANIQVDPSGQVQWLQQQLVPQLQQPLSQQPQQTFPQPQQTFPHQPQQQVP

QPQQPQQPFLQPQQPFPQQPQQPFPQTQQPQQPFPQQPQQPFPQTQQPQQPFPQQPQQPFPQTQQ

PQQPFPQLQQPQQPFPQPQQQLPQPQQPQQSFPQQQRPFIQPSLQQQLNPCKNILLQQSKPASLVSSL

WSIIWPQSDCQVMRQQCCQQLAQIPQQLQCAAIHSVVHSIIMQQQQQQQQQGIDIFLPLSQHEQVGQ

GSLVQGQGIIQPQQPAQLEAIRSLVLQTLPSMCNVYVPPECSIMRAPFASIVAGIGGQ

FIGURE 3.1 **Wheat γ-gliadin sequence**. The sequence is shown in the one-letter code for amino acids, with the proline (P) and glutamine (Q) residues colored blue and red, respectively, and glutamate (E) shown in green.

shows that they make up over a third of the amino acids in the protein (24% **Q**; 11% **P**).

Why could this be problematic? First, as Table 3.1 shows, a "typical" animal protein is balanced in composition, which means it contains a mixed sequence with varied amounts of each of the 20 amino acids.

An average vertebrate protein would not contain more than 3–5% of P or Q. Human tropomyosin (α-chain), a major muscle protein, contains <5% Q and no proline at all!

Secondly, aggregated proteins can be toxic to cells in culture.[2] These sticky proteins (and fragments thereof) may cause problems in the digestive track and may even induce other essential proteins to precipitate with them.

TABLE 3.1

Comparison of Wheat Glutens with Allergens*, Human Tropomyosin and Vertebrate Proteins**, showing (boxed area) they contain less essential amino acids, threonine (T), methionine (M) and lysine (K), and 5-10x more glutamine (Q) and proline (P).

Amino acid:	Ara h 1	Ara h 2	Ara h 3	Ana o 2	Ber e 1	Jug r 1	Pru du 6	Wheat glutenin	Wheat γ-gliadin	Vertebrate Average	Human tropomyosin
Essential:											
His (H)	2.9	1.9	2.1	2.4	2.1	1.4	1.9	0.5	1.6	2.9	0.7
Ile (I)	3.6	1.9	4.1	4.8	2.1	4.3	4.1	0.5	6.1	3.8	4.2
Leu (L)	7.1	10.8	6.7	8.3	6.2	7.2	6.6	4.7	8.1	7.6	11.6
Lys (K)	5.0	1.3	5.2	3.9	0.7	0.0	0.9	0.8	1.0	7.2	13.4
Met (M)	1.5	1.9	1.3	1.3	15.1	2.9	0.2	0.4	1.2	1.8	2.1
Phe (F)	3.6	1.9	3.6	3.7	1.4	2.9	4.3	0.4	4.7	4.0	0.4
Thr (T)	3.7	0.6	3.3	4.2	4.1	2.9	2.3	2.9	2.9	6.2	2.8
Trp (W)	1.0	0.6	0.8	1.5	0	0.7	0.6	1.1	0.7	1.3	0.0
Val (V)	5.3	1.3	5.2	7.0	3.4	3.6	5.1	2.5	5.2	6.8	3.2
Non-essential											
Ala (A)	5.7	6.4	5.6	5.5	7.5	5.8	6.4	3.5	5.5	7.4	12.7
Asn (N)	5.8	4.5	6.5	5.7	2.1	4.3	7.9	0.0	4.8	4.4	1.8
Asp (D)	4.8	6.4	5.1	5.5	0.7	5.0	2.6	0.5	1.6	5.9	8.5
Glu (E)	11.3	8.9	11.0	7.9	12.3	9.4	8.1	1.8	4.3	5.8	19.7
Conditionally essential:											
Arg (R)	9.9	12.1	10.1	8.3	11.0	12.9	9.2	1.3	5.0	4.2	5.3
Cys (C)	1.1	5.1	1.1	1.8	5.5	5.8	0.8	0.6	1.4	3.3	0.4
Gln (Q)	6.8	14.6	7.2	8.3	8.9	16.5	19.6	35.3	24.2	3.7	4.9
Gly (G)	6.9	4.5	6.9	7.4	5.5	5.8	8.5	19.6	4.5	7.4	1.1
Pro (P)	6.0	5.1	5.9	4.4	4.8	1.4	4.0	12.9	10.7	2.9	0.0
Ser (S)	6.8	7.0	6.7	6.1	5.5	5.0	4.9	5.5	5.2	8.1	5.3
Tyr (Y)	1.3	3.2	1.5	2.0	1.4	2.2	2.1	5.4	1.2	3.3	2.1

* See Chapter 4, major nut allergens are Ara h 1, 2, 3 (peanut), Ana o 2 (cashew), Ber e 1 (brazil nut), Jug r 1 (walnut) and Pru du 6 (almond) .

** Vertebrate average values are taken from http://www.tiem.utk.edu/~gross/bioed/web-modules/aminoacid.htm.

JAKE AND ELWOOD's GUIDE TO NUTRITION

But there is another problem with these gluten proteins. It might be educational to refer to the eating habits of the heroes of the movie "The Blues Brothers" (Figure 3.2). In this classic, Brother Jake orders two barbecued chickens, which he apparently aims to eat all by himself. Brother Elwood, on the other hand, orders a loaf of white bread, toasted.

Now, obviously neither man will eat what is considered a balanced diet, which should include some fruits and vegetables. Jake obviously has no problem getting enough protein from his meal as each chicken will contain

FIGURE 3.2 **The Blues Brothers** guide to nutrition.

about 170 g of protein. But since the average person can only process about 30–40 g of protein at a time, most of what he eats will probably just add calories and enhance his ample belly. In addition, he may get some vitamins from that barbecue sauce and, if he eats the skin as well, the fat-soluble vitamin A. But he will be also getting a lot of salt.

Now Elwood could, in theory, obtain a full daily protein requirement (35–50 g) just from eating half a loaf of white bread a day (15 slices, about 1200 calories) and probably a fair amount of essential vitamins, since it will be made from "enriched" flour.

In comparison to Jake, Elwood would be eating relatively larger quantities of Q and P in his proteins, although it is questionable if his body would have full access to them. Even if the pepsin in his stomach could fully digest the gluten peptides (see below), the bottom lines in Table 3.1 show that P and Q are considered only "conditionally essential". This means that under certain conditions, such as after a severe illness ([3] but see also [4, 5]), supplements containing Q may aid recovery.

Over the long term, his body would be starved for some vitamins and other nutrients. Modern white bread is fortified with vitamins, many of which may be altered by the heat of the toaster. As the boxed area in Table 3.1 illustrates, wheat glutenin and gliadins, which make up 80% of the protein in bread, are lacking in "essential" amino acids, the ones human bodies do not produce. Wheat proteins do not supply much methionine (M), lysine (K) or threonine (T), all essential amino acids. The first three columns of the table show that if Elwood put peanut butter on that bread, he would get a much better balance of amino acids. However, Ara h 1, 2, and 3 are potent allergens, causing reactions in those sensitized to peanuts and tree nuts, as will be further discussed in Chapter 4.

IS GLUTAMINE ESSENTIAL?

Many biologists will be surprised this is even a question, as Q is almost always added to culture media for mammalian cells to grow outside the body. Some pathogens even produce virulence factors to starve cells of Q, as will be discussed in Chapter 8. However, Q is one of the most common amino acids in sera. Evidence for Q as conditionally, rather than non-essential, is rather limited, relying as it does on studies where adding Q to nutritional supplements seemed to help recovery of severely ill patients. Some studies suggested adding Q in such formulations did enhance some parameters of recovery for severely ill patients.[6] Finding enough patients for a controlled, confirmatory study of this question is difficult. Most patients in the ICU are not there for very long (many come in as trauma patients) and few require enteral (feeding tube) or parenteral (IV) nutrition. Combining the results of several more recent studies, using Q-enhanced parenteral nutrition supplements for patients in several ICUs, had more equivocal results.[4] Thus, the role of adding glutamine is currently controversial. Further, glutamine is not very soluble or stable (see Figure 3.3), meaning that adding it will lower the shelf life of supplements.

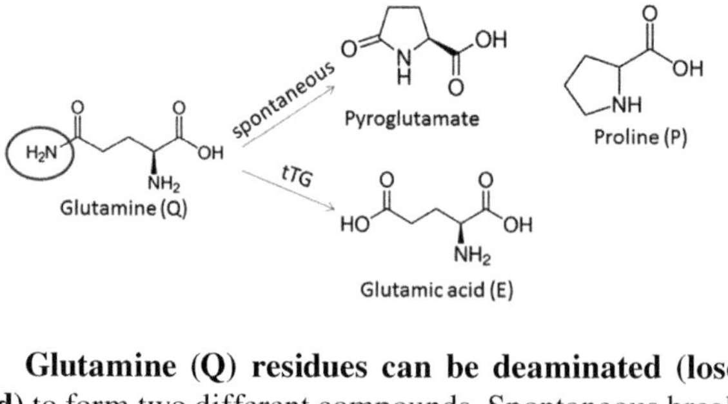

FIGURE 3.3 Glutamine (Q) residues can be deaminated (lose an amino group; circled) to form two different compounds. Spontaneous breakdown, after sitting at room temperature for extended periods, leads to the formation of pyroglutamic acid, which is found at the ends of proteins occasionally and bears a distinct resemblance to proline (P). Tissue transglutaminase (tTG) can convert Q to glutamate (E) residues, making peptides more acidic, soluble and likely to react with T-cell receptors. tTG is normally an intracellular protein; antibodies to its presence in serum are a major indicator of CeD.

Baking makes gluten even harder to digest: As noted above, the gluten in flour/water pastes is sticky and does not dissolve well. Processing wheat proteins to make bread makes them even less soluble. The kneading serves to induce "cross linking", mediated by cysteine (C) residues, which can chemically connect to one another. This means that the proteins become inter-wound, like a ball of yarn after a cat has played with it. This cross linking makes dough elastic, yielding the chewy texture of a good pizza crust. However, gluten in the stomach is not readily available to digestive proteases. A normal protein will get digested in the stomach into short "peptides" containing up to about 15 amino acids. But if one degrades γ-gliadin with pepsin, a protease in the stomach that continues to function in acid, several much longer peptides will result (Figure 3.4):

PHQPQQQVPQPQQPQQP
AMAITIGTANIQVDPSGQVQ
FLQPQQPFPQQPQQPFPQTQQPQQPFPQQPQQPFPQTQQPQQPFPQQPQQPFPQTQQPQQPFPQL
QQPQQPFPQPQQQL
PQPQQPQQSF
PQSDCQVMRQQCCQQ
QCAAIHSVVHSIIMQQQQQQQQQQGIDI
LSQH*E*QVGQGS
VQGQGIIQPQQPAQ
YVPP*E*CSIMRAPFASIVAGIGGQ

FIGURE 3.4 Predicted pepsin cleavage products of γ-gliadin. Note the results of the cleavage are many very long peptides containing long repeat stretches of Q residues. These can form even more indigestible aggregates.

One of those peptides is 28, another 65 amino acids long! These long, insoluble peptides are expected to contribute to gluten-induced CeD. Worse, neither of these long peptides can be further digested by the other proteases they will encounter as they progress down from that acidic stomach into the neutral area of the duodenum. Gluten baked into bread is even less digestible![7] The proteins are also cloaked by lipids (fats) and starch, which have to be removed by other enzymes called lipases and amylases, respectively. Taking medications for esophageal reflux, such as proton pump inhibitors, may reduce the amount of acid in the stomach, making it even more difficult for pepsin to do its' job.

On top of being difficult to digest by proteases, gluten may have a completely different mechanism of toxicity that could be related to the neurological problems that are part of the CeD syndrome.

Converting Q to E with tTG: One possible reason for gluten toxicity for CeD patients may be due to its conversion to another amino acid, glutamate (or glutamic acid, symbol E, as G stands for glycine), which is glutamine minus an amino group (Figure 3.3). As Table 3.1 shows, gluten contains relatively low amounts of glutamate, a non-essential amino acid produced in large amounts by the body. In addition to being relatively insoluble in its free form, glutamine can spontaneously lose its amino group, to become "pyroglutamate" (Figure 3.3), even if stored as a dry powder for long periods of time.

Glutamine (Q) can also lose its amino group and become glutamate (E) by the action of an enzyme, called tissue transglutaminase (tTG), a protein that normally stays near the surface of cells. The sera of CeD patients typically contain high levels of antibodies to tTG. Physicians now take advantage of this fact to do a blood test for CeD, which in many cases can be used to diagnose the disease and start patients on a gluten-free (GF) diet.

This conversion of Q to E in the gliadin peptides completely alters their properties. First, it makes them more acidic, and possibly more soluble. Secondly, conversion of Q to E residues makes them more likely to bind to receptors on a type of blood cell called T-cells, which eventually stimulate B-cells to make antibodies. CeD patients consistently have certain immune receptors on their T-cells (98% of celiac patients have either an HLA-DQ2.5 or -DQ8 receptor, while only 47% of the general population does). This leads to two different paths, a vicious cycle of inflammation in CeD (Figure 3.5).

The cycle begins when gluten is eaten and partially digested by pepsin into insoluble peptides, which can directly damage mucosal epithelial layers and lead to inflammation and release of enzymes such as tTG.[8] The peptides can be converted by tTG to ones that bind better to the T-cells of gluten-sensitive people. This activates T-cells to release cytokines (discussed in more detail in Chapters 5 and 6). The inflammatory response can be further activated by other cytokines released by damaged and inflamed cells. The T-cells can also direct the formation of antibodies by B-cells with specificity for binding to tTG and gluten peptides, which may lead to further inflammation and prevent regeneration of a healthy mucosa.[9,10] Most notably, the tTG enzyme can also induce cross linking of gluten proteins, making them even more insoluble.

PolyQ REPEATS IN PROTEINS THAT AGGREGATE IN NEUROLOGICAL DISEASES

Q residues in those peptides from the pepsin digest of gluten (Figure 3.4) come in groups. The most extreme example, QCAAIHSVVHSIIMQQQQQQQQQQGIDI, contains a 10mer repeat of Q. It contains no prolines and thus would not be further cleaved even by prolyl-peptidase.

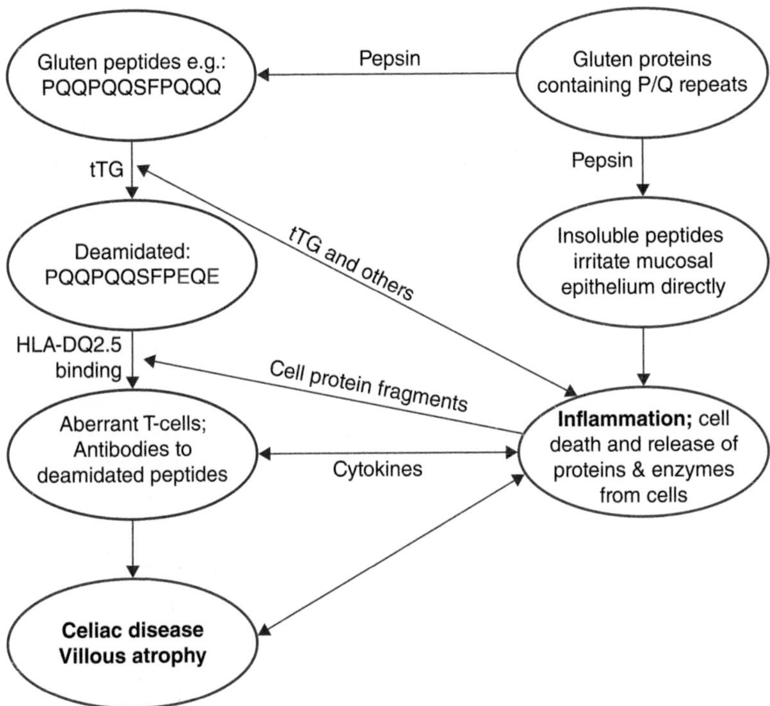

FIGURE 3.5 A vicious circle is proposed for gluten-induced inflammation in CeD. There are two pathways leading to inflammation, a direct one due to insoluble peptides from gluten and their aggregates (red ovals) and an immune controlled one (blue ovals), where soluble peptides are deamidated (Q to E) by increasing amounts of tTG (which can also insert additional chemical bonds between fragments and proteins, making them more insoluble). Peptides with more E residues bind better to some HLA receptors on T-cells, thereby activating the T-cells to make inflammatory cytokines, and to direct the formation of antibodies to gluten and proteins such as tTG released from damaged cells.

Possibly related to this, repeats of Q are also found in human proteins that have been linked to some genetically linked, neurological diseases (Table 3.2).[11,12] Mutations leading to very long repeats of Q (>25–30) have been linked to Huntington disease (HD), which stunted the life and song writing career of Woody Guthrie, among others. In people born with these long repeats, with age and as cells divide, the repeats of Q residues can become progressively longer, causing visible aggregated protein inclusions in the brain that correlate with psychiatric and physical problems. As Table 3.2 shows,

TABLE 3.2
Human Proteins Containing Extended PolyQ Segments, Implicated in Many Neurological Diseases. The numbers in the third column show the position of the amino acid sequence in the indicated proteins.

Human Protein	Disease/Syndrome	PolyQ Region
Androgen receptor polyQ domain	Androgen insensitivity syndrome (AIS)	54LLLLQQQQQQQQQQQQQQ QQQQQQQQQETSPRQQQQQ QGE93
Atrophin 1	Dentatorubral-pallidoluysian atrophy (DRPLA)	478HPPVSTHHHHHQQQQQQQQQ QQQQQQQQQQHH504
Ataxin-1	Spinocerebellar ataxia 1 (SCA1)	194EQQQQQQQQQQQQHQHQQQ QQQQQQ QQQQQHLSRA230
Ataxin-2	Spinocerebellar ataxia 2 (SCA2)/ Amyotrophic lateral sclerosis13 (ALS13)	160MSLKPQQQQQ QQQQQQQQQ Q QQQQQQQQPP190
Ataxin-3	Spinocerebellar ataxia 3 (SCA3)	290KQQQKQQQQQQQQQQGDLS G310
Ataxin-7	Spinocerebellar ataxia 7 (SCA7)	10GEPRRAAAAAGGAAAAAARQ QQQQQQQQPPPPQPQRQQHPP PPPRRTRP60 710 SPLLVHSSSSSSSSSSSSHS730
Voltage-dependent P/Q-type Ca^{2+}channel subunit alpha-1A	Spinocerebellar ataxia 6 (SCA6)	2211HHHHHHHHHHPPPPDKD RYA GPPQQQQQQQQQQQAVARPG 2330
TATA-box-binding protein	Spinocerebellar ataxia 17 (SCA17)	50ILEEQQRQQQQQQQQQQQQQ QQQQQQQQQQQQQQQQQQQQ QQQQQAVAAA100
Huntingtin	Huntington disease (HD)	10FESLKSFQQQQQQQQQQQQQQ QQQQQQQPPPPPPPPPPPQLPQP PPQAQPLLPQPQP PPPPPPPPPPGPAV80

many neural diseases have been directly tied to amplification of long repeats coding for the amino acid glutamine (Q). In those suffering from diseases such as HD, which is a genetic disorder characterized by long repeats of the codon CAG (which encodes the amino acid Q) in the huntingtin protein. Just as the Sorcerer's Apprentice in Fantasia seemed doomed to keep making more and more copies of himself, carrying water until it threatened to drown him, the length of the repeated CAG region in the huntingtin protein gene keeps growing throughout the lifetime of the afflicted.[13] These people develop dementia as they age, sometimes characterized by violent rages and movement disorder. In general, the length of repeat can predict the age of symptom onset. Mechanisms have been suggested to account for how the repeated DNA region can gradually increase its size.[14] The long polyQ segments of the resulting protein may be resistant to degradation or may directly affect the function of another protein involved in autophagy, as discussed elsewhere.[15]

The toxicity of very long "polyQ" segments may be related to their tendency to form insoluble aggregates, which cannot easily be degraded by proteases in cells. Many larger viruses also contain polyQ repeats.[15] One should also note that the translation of repeated codons in the RNA for polyQ and polyP tracts may also affect the ability of ribosomes to translate the encoded proteins, leading to early termination and fragments of the protein that can easily aggregate.[16]

As with the polyQ proteins in the table, fragmented gluten may form toxic aggregates in the intestine. In this, they resemble other proteins, such as those that aggregate in the brain in Alzheimer's disease or block the glomerular capillaries of the kidney. (See Chapter 7, on protein aggregation diseases and the search for compounds or peptides that enhance solubility.[17,18]) However, the blood of HD patients does not contain the antibodies to tTG that are common in CeD patients. This emphasizes that the two diseases are quite distinct. In addition to designing genetic treatments for the polyQ-related diseases listed in Table 3.2, several groups are developing inhibitors of protein aggregates as treatments. While these have been shown to work in cell culture, testing in disease models is ongoing. Even with positive results, it may be many years before they are approved for human use. These inhibitors may eventually lead to therapies for CeD as well.

POSSIBLE TREATMENTS FOR CeD

Elimination diets, in the absence of specific treatments for CeD, may be the only way currently to prevent symptoms. Many question whether research into new treatments for food sensitivities makes sense, as one could simply avoid eating the problematic food. There are now improved GF foods, even pastas and breads, in most supermarkets, making it much easier to adhere to a GF diet. However, maintaining an elimination diet is hard and often lonely work, as discussed in the introduction before this chapter. Further, the intestinal epithelia continue to degrade in many patients and it does not fully recover even in CeD patients following a strict GF diet. Doctors may attribute this to traces of gluten even in GF-labeled foods or accuse the patient of not following the diet carefully enough.

Treatments may come from the trials of drugs that might be repurposed[19] after showing efficacy in other autoimmune conditions (such as oral budesonide and other steroids[20]) or those developed to reduce protein aggregates in neurological diseases. Refractory CeD patients should be tested for other complicating diseases, such as inflammatory bowel syndromes.[21] Psoriasis and dermatitis herpetiformis are often co-morbidities in those with skin manifestations; some refractory CeD patients respond very well to treatment with TNF inhibitors such as infliximab[22] (see also Storming Cytokines, Chapter 6). In addition to glutens, other poorly soluble proteins in grains, such as Amylase trypsin inhibitors (ATIs), may also cause complications; ATIs may be degraded better in sourdough breads.[23]

Controlled trials in refractory CeD patients who have persistent symptoms even when following a strict GF diet have so far failed to yield specific treatments that showed significant symptomatic improvement. It is encouraging that after many years of blaming patients for not following a GF diet, for those who relapse when only a small amount of gluten is consumed, there may be new therapies coming. Even those that could slightly increase the amount of gluten that could be ingested before severe symptoms occur would be welcome.

Better treatments may result from following up on the immunological and protein-based clues of the disease. One suggestion, given the long peptides remaining after pepsin digestion, was that eating a "prolyl-peptidase" with food, i.e., an enzyme cleaving peptides after prolines (P) could be a potential treatment for CeD. However, dosing such a protease presents other problems. First, the protease must be specific for gluten. If not, the enzyme can attach to other proteins in the stomach, distracting it from glutens, and perhaps attacking the proteins that make up the stomach lining. In addition, one needs to eat a lot of prolyl-peptidase. Native pepsin is produced in large amounts and released into the stomach just when it is needed (during eating). How long will the extra protease remain? Will the person be conscientious about swallowing capsules before meals? In addition, the peptides shown in Figure 3.4 assume pepsin (or prolyl-peptidase, if one is going that route) can fully access the protein. This may not be possible in that twisted gluten mass formed after kneading, where there are chemical bonds between chains blocking the enzyme.

One possible treatment route involves inhibiting gluten peptides from leaving the intestines. Partially digested, cross-linked and insoluble, Q-rich peptides may break down the areas between cells (junctions) and move on into other areas of the body. The result is that they can even make deposits in neural tissue and interfere with cognition. This could account for the "brain fog" that many CeD sufferers complain of before they start on a GF diet. A human protein, zonulin, controls tight junction function between intestinal cells[24] and can increase barrier permeability. The small peptide larazotide (GGVLVQPG; Gly-Gly-Val-Leu-Val-Gln-Pro-Gly) was developed as an antagonist of zonulin. Animal and early studies in humans suggested larazotide would prevent gluten peptides from reaching the blood stream by decreasing the porosity of the small intestine. One phase 2 study indicated that larazotide might help in CeD patients following a GF diet. However, a phase 3 study was halted in June of 2022 as the effect was too small to be significant. Still, by reinforcing the tight junctions, larazotide might still be useful in treating other diseases related to intestinal permeability.

CONCLUSIONS

Most people do not rely on dry toast for their proteins but still enjoy eating foods that get their texture from wheat gluten. Glutens have an unbalanced amino acid composition, containing long stretches of proline and glutamine (Ps and Qs) that are also found in human proteins that aggregate in neurological diseases and ataxias. Glutens cause problems for a small percentage of the population for reasons that are not completely resolved. Gluten is not easily digested, yielding long peptides that can activate an immune response, and they and their peptide digestion products aggregate easily. Antibodies to certain proteins in the blood can replace a traditional biopsy for diagnosis of CeD.

One solution for those with sensitivity is to find alternatives to gluten with similar texture. The many gluten-free foods that incorporate proteins from other plants are a tribute to this effort. Additional therapies may include consuming prolyl-peptidases to aid in digesting gluten and enhancing the intestinal barrier (possibly using other zonalin inhibitors than larazotide) to prevent toxic peptides generated from gliadin from entering sera. While none of the CeD-specific therapies tested so far could, on its own, provide significant healing power beyond that of a GF diet, better genetic analysis may reveal new gene targets. Until there is a better therapy for most CeD patients, all people should understand the problems of those trying to adhere to an elimination diet and help accommodate them.

REFERENCES

1. Wieser, H. Chemistry of gluten proteins. *Food Microbiol* **24**, 115–119 (2007).
2. Schein, C.H. Solubility as a function of protein structure and solvent components. *Bio/Technology* **8**, 308–317 (1990).
3. Jones, C., Palmer, T.E. & Griffiths, R.D. Randomized clinical outcome study of critically ill patients given glutamine-supplemented enteral nutrition. *Nutrition* **15**, 108–115 (1999).
4. Mulherin, D.W. & Sacks, G.S. Uncertainty about the safety of supplemental glutamine: an editorial on "A randomized trial of glutamine and antioxidants in critically ill patients". *Hepatobiliary Surg Nutr* **4**, 76–79 (2015).
5. Sacks, G.S. Effect of glutamine-supplemented parenteral nutrition on mortality in critically ill patients. *Nutr Clin Pract* **26**, 44–47 (2011).
6. Griffiths, R.D. Glutamine: the struggle for proof? *Crit Care Med* **39**, 1546–1547 (2011).
7. Smith, F. et al. Digestibility of gluten proteins is reduced by baking and enhanced by starch digestion. *Mol Nutr Food Res* **59**, 2034–2043 (2015).
8. Koning, F. Adverse effects of wheat gluten. *Ann Nutr Metab* **67 Suppl 2**, 8–14 (2015).
9. Al-Bawardy, B. et al. Celiac disease: a clinical review. *Abdom Radiol (NY)* Abdom Radiol (NY). 2017 Feb;42(2):351-360. doi: 10.1007/s00261-016-1034-y.
10. Klock, C., Diraimondo, T.R. & Khosla, C. Role of transglutaminase 2 in celiac disease pathogenesis. *Semin Immunopathol* **34**, 513–522 (2012).
11. Onodera, O. et al. Oligomerization of expanded-polyglutamine domain fluorescent fusion proteins in cultured mammalian cells. *Biochem Biophys Res Commun* **238**, 599–605 (1997).
12. Margolis, R.L. & Rudnicki, D.D. Pathogenic insights from Huntington's disease-like 2 and other Huntington's disease genocopies. *Curr Opin Neurol* **29**, 743–748 (2016).
13. Ciosi, M. et al. A genetic association study of glutamine-encoding DNA sequence structures, somatic CAG expansion, and DNA repair gene variants, with Huntington disease clinical outcomes. *EBioMedicine* **48**, 568–580 (2019).
14. Franklin, A., Steele, E.J. & Lindley, R.A. A proposed reverse transcription mechanism for (CAG)n and similar expandable repeats that cause neurological and other diseases. *Heliyon* **6**, e03258 (2020).
15. Schein, C.H. Polyglutamine repeats in viruses. *Mol Neurobiol* **56**, 3664–3675 (2019).
16. Aviner, R. et al. Ribotoxic collisions on CAG expansions disrupt proteostasis and stress responses in Huntington's Disease. *bioRxiv*, Aviner, R. et al doi: https://doi.org/10.1101/2022.05.04.490528.
17. Chen, D., Martin, Z.S., Soto, C. & Schein, C.H. Computational selection of inhibitors of Abeta aggregation and neuronal toxicity. *Bioorg Med Chem* **17**, 5189–5197 (2009).
18. Bruce, N.J. et al. Molecular dynamics simulations of Abeta fibril interactions with beta-sheet breaker peptides. *Peptides* **31**, 2100–2108 (2010).
19. Schein, C.H. Repurposing approved drugs on the pathway to novel therapies. *Med Res Rev* **40**, 586–605 (2020).

20. Hujoel, I.A. & Murray, J.A. Refractory celiac disease. *Curr Gastroenterol Rep* **22**, 18 (2020).
21. Bramuzzo, M. et al. Phenotype and natural history of children with coexistent inflammatory bowel disease and celiac disease. *Inflamm Bowel Dis* **27**, 1881–1888 (2021).
22. Rawal, N., Twaddell, W., Fasano, A., Blanchard, S. & Safta, A. Remission of refractory celiac disease with infliximab in a pediatric patient. *ACG Case Rep J* **2**, 121–123 (2015).
23. Huang, X. et al. Sourdough fermentation degrades wheat alpha-amylase/trypsin inhibitor (ATI) and reduces pro-inflammatory activity. *Foods* **9**, 943 (2020).
24. Slifer, Z.M., Krishnan, B.R., Madan, J. & Blikslager, A.T. Larazotide acetate: a pharmacological peptide approach to tight junction regulation. *Am J Physiol Gastrointest Liver Physiol* **320**, G983–G989 (2021).

4 The Peanut Gallery and Other Allergen Families

In the etiology of hay fever, then, these three elements prevail – a nervous constitution, an irritable nasal mucosa, and the stimulus .... There is an association with the presence of pollen in the atmosphere, but this is only one of a host of exciting causes.

William Osler, Autumnal Catarrh (Hay fever) in *Principles and practices of medicine* (1st edition, 1892 by D. Appleton & Company, p. 478–479)

OVERVIEW

1. Avoiding allergenic proteins is very difficult, as they are in the air (pollen, dander, insect residue) and common food stuffs.
2. Food allergies, mostly against milk, eggs, peanuts, soy or wheat affect up to 8% of infants and young children.
3. Even tiny amounts of certain foods can cause anaphylactic shock in sensitized individuals. However, defining the exact sensitivities of patients is challenging.
4. As with glutens, most of the population consumes known allergens every day with no harmful effects.
5. Unlike glutens, allergenic proteins resemble other proteins in their amino acid content and can come from very diverse plant and animal sources. There is no one common allergen.
6. Allergenic proteins belong to protein families (PFAM) as indicated by commonalities in their amino acid sequences, 3D-structures (i.e., how the protein folds) and function.
7. Distinct areas of allergens, called epitopes, have been defined that mediate binding to IgE antibodies. They can have similar physicochemical properties (PCPs) to areas in proteins with little overall sequence identity.
8. Different forms of immunotherapies, from shots to sublingual, have varying success in controlling symptoms.
9. Designing hypoallergenic plants requires changing many individual proteins.

DOI: 10.1201/9781003333319-5

INTRODUCTION

Allergens are even more difficult to avoid than the gluten proteins discussed in Chapter 3. They float in the air we breathe and lurk in our most common food stuffs (milk, eggs, peanuts, soy or wheat). Pollens from cedar and pine cause misery for many people in Texas and other areas of the United States. In Japan, giant red cedar trees (Sugi, *Cryptomeria japonica*) were originally revered as the national tree and are now reviled as an omnipresent allergy stimulant in the spring months.[1] Further, sensitization from airborne sources can correlate with food allergies, as can be seen by regional differences in sensitivities.[2] For example, those allergic to the pollen of birch trees, found all over northern Europe, cannot eat hazelnuts but may be able to tolerate peanuts. Peanut allergies are found in up to 3% of US children, but hazelnut or almond allergies are rare. Sesame allergies are common in Israel, where tahini and halva are eaten regularly after weaning, while peanut allergy may be held in check by giving babies corn puffs coated with peanut flour ("Bamba").

Allergy: an immune-mediated disease. Allergy symptoms can range from sniffles and sneezes to major asthma attacks, from mild skin irritation to rash and swelling throughout the body. In highly sensitized individuals, exposure can even induce lethal anaphylaxis. Before mid-20th century, noted doctors attributed "hay fever", "rose fever" or "autumnal catarrh" to a nervous disposition in the patient. Physicians were trained to recognize diseases caused by infections, cancers and metabolic errors, such as diabetes. Without a clear understanding of the immune system, they had no way of diagnosing allergic syndromes and often resorted to blaming the patient for any discomfort they were feeling. Some related the accompanying runny nose to nasal disfunction (cauterization of the nostrils was one remedy), while others cast doubt on whether pollen was truly a stimulus. The word allergy (*Allergie* in German, from Greek, *allos*, meaning strange, and *ergon*, activity) did not even exist until it was first coined by the Austrian pediatrician Clemens Freiherr von Pirquet around 1906. We now know the word is remarkably correct: a sensitized immune system recognizes (makes IgE antibodies against) certain otherwise innocuous substances to be foreign, toxic invaders (Figure 4.1). Von Pirquet, an experienced microbiologist, recognized that the delayed and often severe reactions to second injections of a horse serum that was used to treat rheumatic fever (caused by group A *Streptococci*, see Chapter 8) or vaccines, such as those against diphtheria or tuberculosis, were different than those caused by infections and must be due to some process in the body.

Still, the idea that allergy could be due to exposure to pollens or common, nutritious foods took many decades to be universally acknowledged.

Many have noted that it was difficult to implicate cigarettes as a cause of cancer or emphysema when most people smoked, including a large proportion of the medical profession. Doctors even recommended smoking to their patients as a way to "calm your nerves". In the same manner, recognizing allergy to specific foods that most people eat without problems can be very difficult. Unlike the noxious fumes of cigarettes that cause general cell damage, allergenic proteins are only conditionally toxic, as much of their negative effects are mediated by the immune system, particularly through allergen/IgE/cell receptor complex mediated activation of basophils (see Figure 4.1).

Around the same time, doctors found that one could induce what is now referred to as tolerance by repeated injections of extracts of allergenic triggers, such as ragweed, into patients with especially severe symptoms. While these raw extracts relieved symptoms in some patients, the injections could also cause dangerous and occasionally fatal reactions. Such

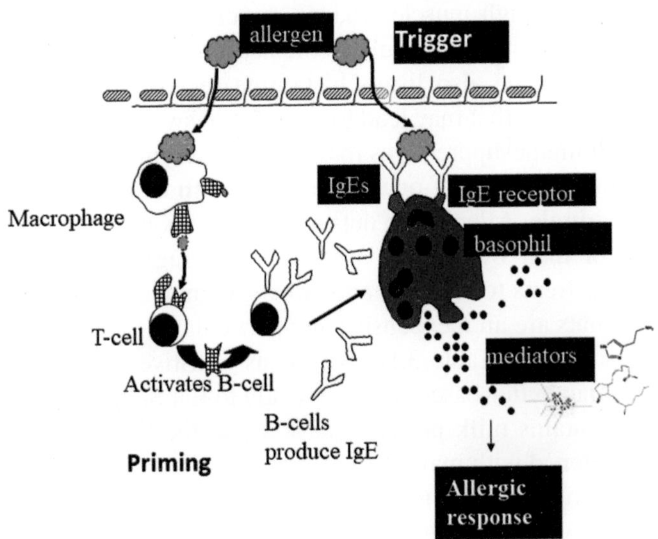

FIGURE 4.1 Allergens can both prime (left side) and then trigger the allergic response (right). Due to cross reactivity, different related allergens can be sensitizers or triggers. In this simplified figure, a sensitizing allergen is consumed by macrophages, which present it (in degraded form) to T-cells, thereby activating them. The activated T-cells in turn stimulate B-cells to produce IgE molecules, which bind to their FcεR1 receptor on basophils. The basophils are now primed to bind the triggering allergen. When the allergenic protein binds to IgE/FcεR1, the basophil releases mediators (histamine, prostaglandins, lipids and other molecules, as well as inflammatory proteins) that affect other cells and start the allergic response.

hypersensitivity reactions had been seen before with blood transfusions, as described in Chapter 2. Extreme responses in some individuals to "infusions" that protected most children against infectious diseases brought the problem into focus. For example, the young son of Dr. Langerhans, for whom the pancreatic islets (where insulin is stored) are named, died suddenly after being injected with an early diphtheria vaccine. This severe form of allergic response was named anaphylaxis (a combination of the Greek words *ana,* meaning up and *phylax,* guardian, protector) or anaphylactic shock.[3] The meaning was clear: the normally protective immune response, when upregulated, could be toxic.

The most dangerous aspects of anaphylaxis are hypotension (a rapid drop in blood pressure) and breathing difficulties. Swallowing an antihistamine tablet is not possible due to throat swelling and would anyway not help fast enough. Instead, the reaction can usually be stopped by rapid injection of the adrenal hormone, epinephrine (adrenaline), a tyrosine derivative first isolated around 1900. Epinephrine, via needle or "Epipen" injection into the thigh muscle, relaxes muscles in the lungs responsible for bronchospasms; it also stimulates the heart to increase blood flow and pressure. Epinephrine also limits histamine release and calms muscles throughout the body that may lead to rash, facial swelling and other symptoms. Other immunosuppressives may also be used in emergency rooms.

Even a small amount of peanut in a food can be dangerous for sensitized individuals. Allergies to nuts are seen in about 3% of US children. As peanuts or their byproducts are present throughout our food chain, from ballgame treats to stir fries and sweets, complete elimination is very difficult. Peanuts are an inexpensive and particularly rich source of essential amino acids (see Table 3.1), as well as nutritive fats, carbohydrates and fiber. Peanut butter-based RUSFs (Ready to Eat Supplementary Foods, which also contains milk proteins) have saved the lives of thousands of children threatened with malnutrition.

Although food allergies in children, especially those to eggs and milk, may be outgrown with time, some are life long and even become more virulent with age, such as that to shrimp or other crustaceans and cedar pollen allergy.[1] Recent results indicate that young children with mild peanut allergy (as determined by the wheal size after a skin prick test) may be desensitized by feeding small amounts of peanut protein (for example letting toddlers suck on peanut puffs). However, this is not recommended if the wheal size is greater than about 1/8 of an inch, as even a small amount of peanut may induce symptoms. Those with more severe allergy are willing to undergo long and potentially dangerous (due to possibly inducing anaphylaxis) immunotherapy treatments to reduce their sensitivity. Desensitization may be done with injections (used with crude extracts for over 100 years) or newer oral (OIT) or sublingual (SIT) ones, where extracts of the allergen are injected under the tongue.[4] However, the process can require years of regular treatment in a doctor's office, is not without danger (there must be ready access to a hospital in case the therapy induces anaphylaxis[4a]) and does not work for many individuals. Treatment success may be measured in how many parts or individual whole peanuts can be consumed without response.

There is hope understanding the special characteristics of allergenic proteins and the epitopes that bind IgE may lead to better and more specific treatments. This may also help design hypoallergenic foods.

ALLERGENS ARE DIFFICULT TO DISTINGUISH FROM OTHER PROTEINS

Unlike glutens, the major allergens of peanuts have amino acid compositions similar to other proteins (Table 3.1) and do not have a single common structure. There are at least 18 different proteins in peanuts (Figure 4.2) that cause allergic response, many of which have can be grouped according to

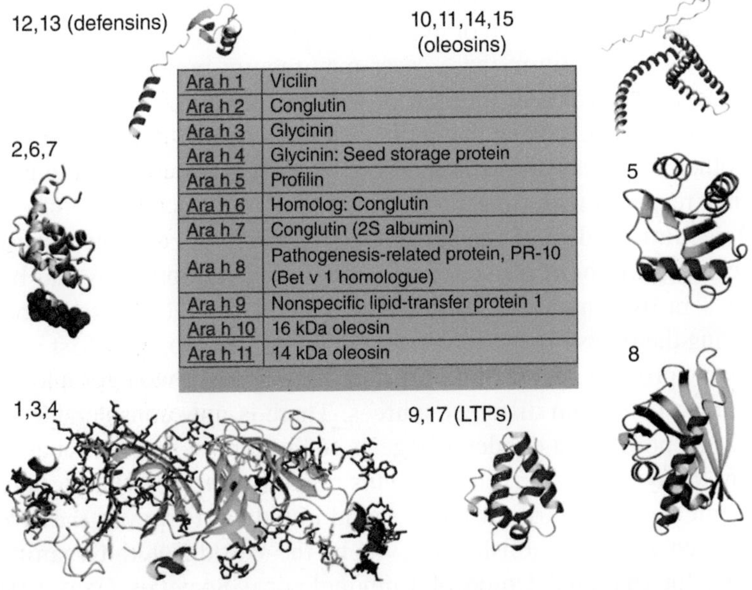

FIGURE 4.2 The diverse structures of the allergenic proteins (Ara h 1-17) of peanut are illustrated. Homologues of these proteins in tree nuts have very similar folded structures, although they come from different botanic sources. The highlighted red area of the similar structure of allergens Ara h 2,6,7 indicates a common IgE epitope identified in the three peanut proteins.

sequences and structural similarity. In addition, similar proteins can occur in many different sources, as discussed below.

The need to identify proteins that were similar to allergens, to avoid inadvertently incorporating them into food and other products people would come into contact with, was indicated by efforts to transfer genes for nutritious proteins from one food to another. One of the first demonstrations that individual proteins could cause allergenic reactions came from the laudable project to introduce a highly nutritious protein, now called Ber e 1, from Brazil nuts into grains. Its' sequence is high in the essential sulfur containing amino acids (see Table 3.1) methionine (Met, M) and cysteine (Cys or C) that are lacking in many plant proteins. Several research groups successfully inserted this nut gene into grains, soybeans and other food crops. However, the promising project was abandoned when it was found that the transgenic soybeans induced a rash in patients who were allergic to Brazil nuts.[5] Further, Ber e 1 was similar to potent allergenic proteins in peanuts, tree nuts, soy, legumes and other foods known to elicit clinically significant responses.[6]

Although major allergens can be grouped using bioinformatic methods, as described below, even therapeutic proteins can sometimes sensitize on repeated injection. Recombinant human insulin, in some rare diabetic patients, may induce a rash or even an anaphylactic response.[7] Continuing treatment with insulin in these cases may require inhibiting the immune response with steroids or interfering with the ability of IgE to bind to its receptor (Figure 4.1) using the antibody, Omalizumab. Antibodies used to treat other illnesses (some discussed in later chapters) can also induce an immune response with time.

Molecular studies of the proteins in extracts of complex allergenic sources began about 60 years ago. Identifying the proteins that bound IgE in sera from hypersensitized individuals led to modern tests for categorizing the types of allergens a patient might react to.[8] The first isolated allergenic proteins, Amb a 1 and Amb a 2 of ragweed[9] were soon followed by many others from different sources. There is now considerable information on pollen and dander allergens, such as those from grass and tree pollen. Many of the allergenic proteins in foods have been better defined, and some have similarity to those found in pollen. However, the systematic nomenclature, established in 1984 by the World Health Organization and the International Union of Immunological Societies (WHO/IUIS), makes it extremely difficult to distinguish common properties of allergenic proteins from name alone. The names of individual isolated allergenic proteins are based on the Latin designation of their sources and the order they were identified. For peanut (Latin species name *Arachis hypogaea*), the first identified major allergen is called Ara h 1, and the second,

Ara h 2. As many as 30% of those allergic to peanut will also react to English walnut (*Juglans regia*), whose allergens are called Jug r 1 and 2, in order of their discovery. But Jug r 1 is closest in properties to Ara h 2, and Jug r 2 is most similar to Ara h 1, which is turn has regions identical to the later discovered peanut allergens Ara h 3 and 4. It gets worse. Two later discovered peanut allergens, Ara h 6, and Ara h 7, cross-react and share large blocks of sequence similarity with Ara h 2.[6] Then a separate, "11S albumin" family of allergens includes proteins Ara h 3 and 4, Pru du 6 (of almond), Ana o 2 (cashew), Pis v 5 and Pis v 2 of pistachio and Cor a 9 of hazelnut. Even someone with a great knowledge of plant biology or protein chemistry would have trouble determining any functional relationship of one allergen to another, given only their names.

Those studying allergy realized that only a computer could keep all the details of these proteins straight.[10] Identifying such details for all possible allergenic proteins is a hefty undertaking but important for those wanting to pinpoint the binding sites (epitopes) for IgE on their surfaces, which requires detailed molecular comparison. The Structural database of allergenic proteins (SDAP) (https://fermi.utmb.edu/) is now over 20 years old[10] (Figure 4.3). The database allows one to find allergenic proteins identified for a given food, plant, dander, insect venom etc. The protein name, type, source and protein family (PFAM) designation of each allergen can be used to find other, possibly non-allergenic relatives. Other tools are designed to see whether a newly discovered protein has regions of similarity to a

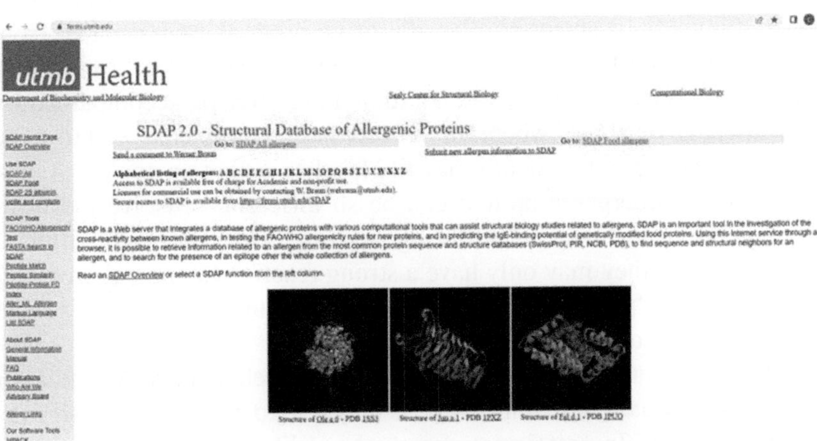

FIGURE 4.3 Home page of the Structural database of allergenic proteins (SDAP), a cross-referenced database of allergenic proteins including data on their sequences, structures, epitopes and functions. Finding relationships between allergens can help define regions responsible for cross-reactivities.

known allergen.[11] These predictions can then be tested experimentally by attaching long linear peptides to a membrane or a plastic surface and measuring the degree to which each binds to IgE in patient sera.[6,12]

Potential cross-reactivities may complicate avoidance diets for those allergic to peanut. One of the most significant uses of SDAP is to find allergens in other sources that resemble known triggers. This is extremely important, as reactions and even death can occur when an individual unknowingly becomes exposed to peanuts or nuts containing similar allergenic proteins in a public place or school. One example, of a seven-year-old Virginia school child, who experienced a severe reaction and anaphylaxis to an unknown source of nut proteins,[13] focused public attention on the severity of reactions to nuts, and the need to develop better preventive treatments for allergy.

We now know that the majority of allergenic proteins can be grouped to only to about 20 different protein families (or Pfams) meaning they have the same shape and share some sequence similarity. Ara h 1 of peanut belongs to the cupin PFAM, a very large family with members that occur throughout the plant kingdom. There are probably hundreds of thousands of different cupins, many (probably most) of them being non-allergenic.

There are many potent allergens, perhaps lurking in a completely different form than the original known source of a protein. In addition to avoiding the foods that sufferers know they react to they may also have to avoid other foods that contain similar proteins. Proteins similar in structure to Jun a 3 (Figure 4A) have been isolated from many fruits and vegetables, which bear no resemblance at all to the cedar tree itself or its pollen. Figure 4.4B shows similar allergenic proteins were isolated from cedar pollen, Jun a 3,[14,15] cherries, apples and green pepper. Cedar pollen allergy typically develops in adulthood and may be accompanied by "oral allergy syndrome" where the sufferer begins to react as well to some foods. For example, the mouth may tingle when eating an apple.[16] As the sources of an allergenic protein can be so different, sensitive individuals may indeed feel they are suffering from "multiple allergy syndrome", when in reality they may only have a strong reaction to one basic protein type. The term "panallergen" was coined for proteins that are present in many different sources.[17]

Dust can sensitize children to dander from household pets, dust mites or cockroach residue. These sensitizations may lead to inability to eat shrimp, crawfish or crabs. Insects do not often find a (deliberate) place in European or American diets, yet their allergenic proteins, including tropomyosin[18] and a fatty acid binding protein,[18] are similar to those of shellfish. This relationship is well known to Louisianans, who refer to crawfish as "mudbugs". Given the number of individuals sensitive to shrimp, crawfish, dust

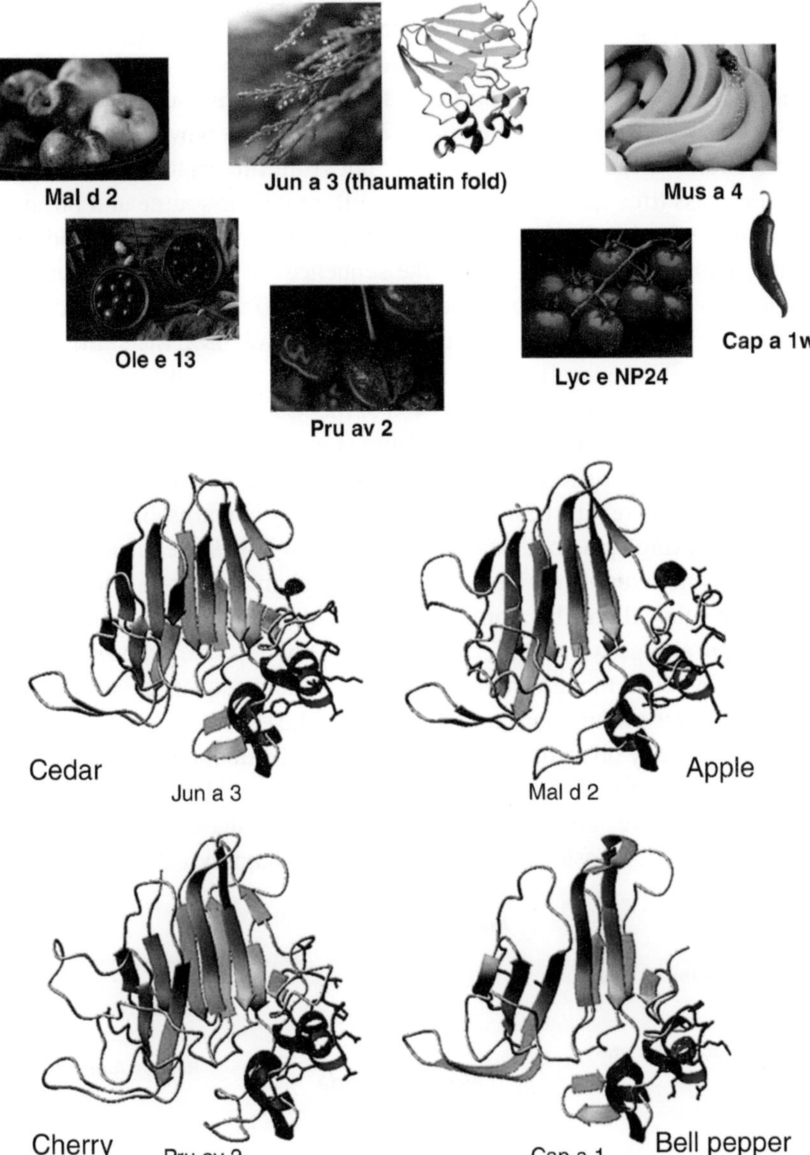

FIGURE 4.4 Allergens with similar structures are found in pollen and foods. (A) After Jun a 3 was identified as a potent allergen in cedar pollen, and shown to have high homology to the sweet protein thaumatin, similar proteins in cherry, apple, bell pepper (B), kiwi and tomato were also found to be allergenic. This suggests that someone sensitized to cedar pollen might develop a sensitivity to these or other foods, as they all contain a similar protein.

mites and cockroach allergens, incorporating ground insects into new meat substitutes may induce reactions in some part of the population.

Some cross-reactivities are particularly common: ~35% of patients who are allergic to peanuts also react to tree nuts, particularly walnuts. Cross reactivity between peanuts and tree nuts may not be surprising, as we are used to seeing mixtures of these in coffee table bowls. The German name Erdnuss, means, appropriately, "nut from the earth", while walnut is Baumnuss (tree nut). No matter how different their source and physical form, these foods contain similar storage proteins making up a good deal of the mass of their seeds. While the sequences of these proteins are not completely identical, the long chain (of about 500 amino acids) folds in a similar way, so the proteins all, at the atomic level, have the same shape, or structure. Those familiar with the evolution of plants will say that major allergens arose long before the division of the Magnoliopsida kingdom into all the different families they come from (Figure 4.5).

No matter how different their physical form, the edible portion of nuts (perhaps in this regard best thought of as seeds) contains similar storage proteins making up a good deal of their mass. While the sequence of these proteins is not completely identical, the long chain folds in a similar way, so the proteins all, at the atomic level, have the same shape, or structure. The major allergens in their edible portions, their seeds, are very similar structurally, even though their overall amino acid sequences are diverse, reflecting their evolutionary distance (Figure 4.5).

Since so many proteins have a similar fold, why do the peanut and some nut proteins specifically evoke such severe responses? Perhaps it would be more constructive, given that the toxin ricin also shares the same cupin fold as Ara h 1, to ask why these proteins are happily consumed by the

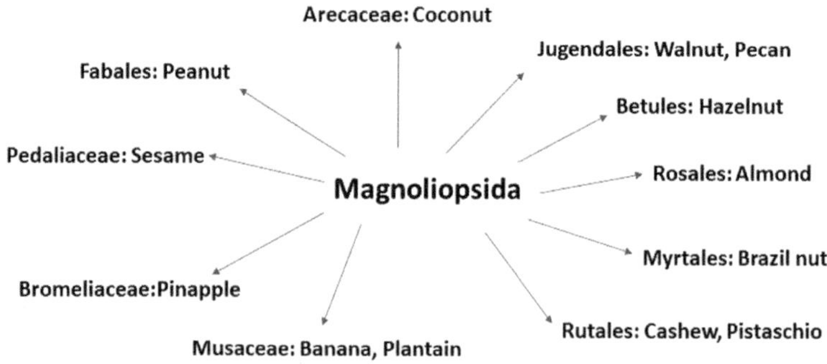

FIGURE 4.5 Allergenic proteins come from the many different plant families of the class Magnoliopsida. While the plants look completely different, they may contain similar proteins (2S and 11S albumins, vicilins, profilins etc.) in their pollen or seeds, fruits or nuts, that sensitize and/or cause cross reactions.

general public and only exert a negative effect on those sensitized to peanuts or tree nuts? It is possible that the immune response was previously triggered by exposure to somewhat less toxic homologues produced by other plants or even bacteria,[19] which have similar structures and mode of action. The sensitized person might have had indigestion or some other symptoms from that encounter, now long forgotten, before having an acute response to the offending food. The adage "what does not kill you makes you stronger" may not apply to the development of allergies.

Cross reactivity is often related to a few highly conserved proteins found in many sources, such as profilins or lipid transfer proteins (LTPs)[20] of plants. The sequences of both of these proteins have been conserved throughout evolution in the kingdom Magnoliopsida (Figure 4.5). For example, the profilins of birch tree pollen (Bet v 2) and grass pollen (Phl p 12) are cross reactive[21] and have 80% identical sequences (identical positions are underlined):

```
Phl p 12    MSWQTYVDEHLMCEIEG--HHLASAAILGHDGTVWAQSADFPQFKPEEI
Bet v 2     MSWQTYVDEHLMCDIDGQASNSLASAIVGHDGSVWAQSSSFPQFKPQEI

Phl p 12    TGIMKDFDEPGHLAPTGMFVAGAKYMVIQGEPGRVIRGKKGAGGITIKKT
Bet v 2     TGIMKDFEEPGHLAPTGLHLGGIKYMVIQGEAGAVIRGKKGSGGITIKKT

Phl p 12    GQALVVGIYDEPMTPGQCNMVVERLGDYLVEQGM
Bet v 2     GQALVFGIYEEPVTPGQCNMVVERLGDYLIDQGL
```

If the sequences were ranked according to their similarity in physicochemical properties (PCPs), the relationship would be close to 100% (which probably means that these proteins serve a vital function in the plants they come from). In the same fashion, cashew and pistachio cross-react, which can be related to the high degree of identity between their 2S albumin sequences.[6] This makes it very important to classify motifs and epitopes in the related proteins that correlate with allergenic potential.[11]

Another distinction can be made between the proteins that induce severe, as opposed to milder allergic symptoms. For example, unlike peanut allergic northern European and US individuals, whose IgE binds predominantly to seed storage proteins (2s, 7s and 11s albumins), IgE in sera of Italians, depending on region, predominantly bind the lipid transport (LTP) allergenic protein, Ara h 17.[22] Other foods common in the Italian diet, including apples, tomatoes, corn and even wine, contain LTP allergens very similar to Ara h 17. Thus, peanut allergy in Italians could be secondary, with initial sensitization from another food. However, the small group of those who suffered the most severe symptoms had IgE reactive to the same seed proteins, especially Ara h 2 and Ara h 6, most detected by IgE in the sera of US and European peanut sensitive individuals.[6]

The need for new treatments for allergy. Those with severe allergies to nut or peanut live in continuous fear of accidental exposure to nuts. The usual symptoms, stomach upset, coughing and sneezing, can be dealt with by antihistamines (which block one or more of the major receptors on cells that bind the inflammatory mediator histamine (see Figure 4.1). However, in many individuals the response can be more severe, leading to skin rashes, tissue swelling and difficulty breathing. This can progress to anaphylactic shock, which can have deadly consequences. Deaths can occur when an individual unknowingly becomes exposed to peanuts (or other sources of similar allergenic proteins) in a public place or school. One example, of a seven-year-old Virginia school child, who experienced a severe reaction and anaphylaxis to an unknown source of nut proteins,[13] illustrates the severity of reactions, and the need to develop better preventive treatments for allergy.

In addition, the need to maintain a nut free food system and atmosphere affects the food industry as a whole. Failure to properly label food that contains even trace amounts of nuts can lead to fines, and even prison time in egregious cases.

Since we know what an allergenic protein "looks like", why not just remove all the allergens from food crops? Recently, there has been talk of producing a hypoallergenic cat,[23] possibly encouraged by the rational design of low IgE binding mutant forms of the Der p 21 of dust mites[24] and Ara h 2 (of peanut).[25] Bioinformatic methods could be used to identify sequences that these related allergens had in common, which could distinguish them from similar proteins that did not evoke an allergic response. Identifying these discrete areas and relating them to the allergens ability to elicit severe IgE-mediated reactions is the first step in our path toward designing our non-allergenic proteins.

This has of course been attempted. But complete removal of a problematic protein in plants is not easy for a number of reasons. Plants have enormous and very repetitive genomes, with some peanut species having 40 chromosomes. A high percentage of plant DNA may be in the form of microsatellites and may contain multiple versions of genes for storage protein such as Ara h 2 or related allergens (see Figure 4.2), making total elimination difficult.[12] Deletion of Ara h 2 did not affect the plant overall, but the peanuts were still allergenic as they still contained allergens such as Ara h 6 and 7. Further, targeting a single protein, as noted above, may not lead to non-allergenic versions of a food. IgE antibodies in the sera of patients with food and pollen allergies typically bind to multiple proteins in those substances; rarely do patients respond to only one isolated allergen. Even small amounts of a related allergen in a food, such as peanut (which contains multiple different allergenic proteins), can still make it unpalatable to allergy sufferers. As noted above, there are similar proteins to Ara h 2, a very potent peanut allergen, in the peanut that have unique and common epitopes.

Further, allergenic proteins are often the most common ones in food sources, meaning removing them would lower their nutritional value. The allergenic proteins in peanuts make up as much as 30% of their mass; 40% of the protein in milk is α-casein, the allergen Bos d 8. Completely removing the major allergen Ara h 1, which makes up 12–16% of total protein, would significantly lower the food quality and strip the peanut of many essential amino acids (Table 3.1). In addition, Ara h 1 and other allergenic proteins can have plant-specific, essential functions, such as protease activity[26–28] or lipid binding.[29] Removing even less abundant proteins can also affect growth. Removing Sol l 1 (profilin, an allergenic protein found in many common foods) from tomatoes reduced IgE binding but lowered the growth rate of the plants. Growth rate could be restored by inserting the gene for yeast profilin,[30] which did not bind IgE from allergic sera, but patients still reacted to other proteins in the tomato extract.

Obviously, the best approach would be to design a new set of peanut proteins, removing or altering the areas that would look like IgE epitopes (i.e., amino acid regions that bind to IgE antibodies in allergic patient sera) while retaining the structure and function of the parent proteins. This may be a difficult proposition, especially for large proteins such as Ara h 1 and its homologues, Ara h 3 and 4, which contain many IgE binding sites. Figure 4.6 shows the surface of Ara h 1, with all the IgE epitopes that have

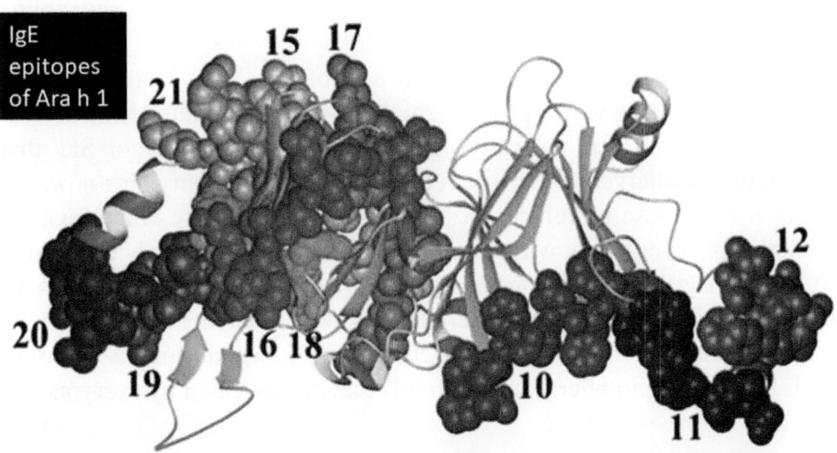

FIGURE 4.6 The IgE epitopes (tone balls) of Ara h1 cover much of its surface. The numbers mark the location of individual peptides found to react with patient IgE. Many of these areas cannot be visualized in the crystal structure, indicating they are highly mobile loops.

been identified marked. Removing all of these areas will only leave us with a skeleton of a protein.

What do epitopes of allergens have in common? We thus have to be smarter and treat our protein with a scalpel rather than a hatchet. Instead of chopping out whole areas, we can change individual amino acids, if we know which points are likely to form the major binding site to antibodies. Just a change of a single amino acid may be enough to eliminate the binding of serum IgE, meaning generating hypoallergenic proteins by site-directed mutagenesis of the plant genome (say with modern Crisper technology[31]) should be possible.

This approach could be preferable to completely removing allergenic proteins, which could affect the ability of the source to grow or reproduce and could greatly lower the nutritional value of foods. The work on identifying IgE epitopes of major allergens is also important for food engineering of artificial meats. If one can identify the areas of a storage protein that are dominant IgE epitopes, one can use this knowledge to design hypoallergenic proteins.

Anticipating allergenicity: conditionally toxic alternative proteins. Although significant advances have been made in designing hypoallergenic peanuts, by selection or recombinant technology, we are still far from having a cultivar that could be called safe.[32] A major effort is also underway to identify cultivars of peanuts (selected from seed libraries) that produce less reactive isoforms of the major peanut allergens or to lower the concentration of two potent peanut allergens, Ara h 2 and Ara h 6, without altering the ability of peanut plants to germinate, their overall nutritional content or ability to resist fungi.

The peanut story covers only one aspect of how useful bioinformatics can be in predicting which proteins might be allergens. Detailed structural data on allergens may aid in designing individual proteins for SIT that have reduced allergenicity while retaining immunogenicity. For now, the major use of these bioinformatics studies is to determine what environmental triggers a patient with a known sensitivity is most likely to respond to. They may thus help to avoid life threatening responses to substances that would otherwise be considered innocuous. But in the future, these tools may help in the design of novel treatments, which directly interfere with IgE binding to allergens, and foods that can be eaten by everyone.

CONCLUSION

Allergens similar to those we already know about can lurk in many different forms. Although the use of the Brazil nut, Ber e 1 protein, to boost the nutritional value of crops had to be abandoned, the idea of incorporating nutritious or useful proteins was not. The caveat emerged that one had to make sure the protein would induce neither a toxic nor an allergenic response. Knowing what allergens look like, in sequence and structure and pinpointing important amino acids for IgE binding, may allow us to identify new, hypoallergenic varieties of the peanuts and other problematic foods. This can also help predict what proteins in other plants might also cause problems for sensitized individuals.

REFERENCES

1. Midoro-Horiuti T, Schein CH, Mathura V, Braun W, Czerwinski EW, Togawa A, Kondo Y, Oka T, Watanabe M, Goldblum RM. Structural basis for epitope sharing between group 1 allergens of cedar pollen. Mol Immunol. 2006;43(6):509–18. Epub 20050621. doi: 10.1016/j.molimm.2005.05.006. PubMed PMID: 15975657; PMCID: PMC2596064.
2. Lipp T, et al. Heterogeneity of pollen food allergy syndrome in seven Southern European countries: the @IT.2020 multicenter study. Allergy. 2021;76(10):3041–52. Epub 20210718. doi: 10.1111/all.14742. PubMed PMID: 33492738.
3. Ring J, Grosber M, Brockow K, Bergmann KC. Anaphylaxis. Chem Immunol Allergy. 2014;100:54–61. Epub 20140515. doi: 10.1159/000358503. PubMed PMID: 24925384.
4. Nagendran S, Patel N, Turner PJ. Oral immunotherapy for food allergy in children: is it worth it? Expert Rev Clin Immunol. 2022;18(4):363–76. Epub 20220316. doi: 10.1080/1744666X.2022.2053675. PubMed PMID: 35285356.

4a. Midoro-Horiuti, T. & Schein C.H. Peptide immunotherapy for aeroallergens. HYPERLINK "https://www.ncbi.nlm.nih.gov/pmc/articles/PMC10362967/" Allergy Asthma Proc. 2023 Jul; 44(4): 237–243. doi: 10.2500/aap.2023.44.230028

5. Nordlee JA, Taylor SL, Townsend JA, Thomas LA, Bush RK. Identification of a Brazil-nut allergen in transgenic soybeans. N Engl J Med. 1996;334(11):688–92. doi: 10.1056/NEJM199603143341103. PubMed PMID: 8594427.
6. Dreskin SC, Koppelman SJ, Andorf S, Nadeau KC, Kalra A, Braun W, Negi SS, Chen X, Schein CH. The importance of the 2S albumins for allergenicity and cross-reactivity of peanuts, tree nuts, and sesame seeds. J Allergy Clin Immunol. 2021;147(4):1154–63. Epub 20201118. doi: 10.1016/j.jaci.2020.11.004. PubMed PMID: 33217410; PMCID: PMC8035160.
7. Aberumand B, Jeimy S. The complexities of insulin allergy: a case and approach. Allergy Asthma Clin Immunol. 2021;17(1):79. Epub 20210729. doi: 10.1186/s13223-021-00554-1. PubMed PMID: 34325725; PMCID: PMC8320219.
8. Johansson SGO. The discovery of IgE. J Allergy Clin Immunol. 2016;137(6):1671–3. Epub 20160423. doi: 10.1016/j.jaci.2016.04.004. PubMed PMID: 27264002.
9. King TP, Norman PS, Lichtenstein LM. Studies on ragweed pollen allergens. V. Ann Allergy. 1967;25(10):541–53. PubMed PMID: 4168086.

10. Schein CH, Negi SS, Braun W. Still SDAPing along: 20 Years of the structural database of allergenic proteins. Front Allergy. 2022;3:863172. Epub 20220322. doi: 10.3389/falgy.2022.863172. PubMed PMID: 35386653; PMCID: PMC8974667.
11. Lu W, Negi SS, Schein CH, Maleki SJ, Hurlburt BK, Braun W. Distinguishing allergens from non-allergenic homologues using physical-chemical property (PCP) motifs. Mol Immunol. 2018;99:1–8. Epub 20180406. doi: 10.1016/j.molimm.2018.03.022. PubMed PMID: 29627609; PMCID: PMC5994374.
12. Nesbit JB, Schein CH, Braun BA, Gipson SAY, Cheng H, Hurlburt BK, Maleki SJ. Epitopes with similar physicochemical properties contribute to cross reactivity between peanut and tree nuts. Mol Immunol. 2020;122:223–31. Epub 20200519. doi: 10.1016/j.molimm.2020.03.017. PubMed PMID: 32442779.
13. Green TD, Palmer KP, Burks AW. Delayed anaphylaxis to walnut following epinephrine administration. J Pediatr. 2006;149(5):733–4. doi: 10.1016/j.jpeds.2006.06.048.
14. Soman KV, Midoro-Horiuti T, Ferreon JC, Goldblum RM, Brooks EG, Kurosky A, Braun W, Schein CH. Homology modeling and characterization of IgE binding epitopes of mountain cedar allergen Jun a 3. Biophys J. 2000;79(3):1601–9. doi: 10.1016/S0006-3495(00)76410-1. PubMed PMID: 10969020; PMCID: PMC1301052.
15. Midoro-Horiuti T, Goldblum RM, Kurosky A, Wood TG, Schein CH, Brooks EG. Molecular cloning of the mountain cedar (Juniperus ashei) pollen major allergen, Jun a 1. J Allergy Clin Immunol. 1999;104(3 Pt 1):613–7. doi: 10.1016/s0091-6749(99)70332-5. PubMed PMID: 10482836.
16. Hofmann A, Burks AW. Pollen food syndrome: update on the allergens. Curr Allergy Asthma Rep. 2008;8(5):413–7. doi: 10.1007/s11882-008-0080-0. PubMed PMID: 18682109.
17. Kleine-Tebbe J, Ackermann-Simon J, Hanf G. Molecular allergy diagnosis using pollen marker allergens and pollen panallergens: five patterns seen in multiple test reactions to pollen extracts. Allergol Select. 2021;5:180–6. Epub 20210527. doi: 10.5414/ALX02238E. PubMed PMID: 34079923; PMCID: PMC8167734.
18. Munera M, Martinez D, Wortmann J, Zakzuk J, Keller W, Caraballo L, Puerta L. Structural and allergenic properties of the fatty acid binding protein from shrimp Litopenaeus vannamei. Allergy. 2022;77(5):1534–44. Epub 20211030. doi: 10.1111/all.15154. PubMed PMID: 34695231.
19. Sandvig K, Kavaliauskiene S, Skotland T. The protein toxins ricin and Shiga toxin as tools to explore cellular mechanisms of internalization and intracellular transport. Toxins (Basel). 2021;13(6). Epub 20210525. doi: 10.3390/toxins13060377. PubMed PMID: 34070659; PMCID: PMC8227415.
20. Di Muzio M, Wildner S, Huber S, Hauser M, Vejvar E, Auzinger W, Regl C, Laimer J, Zennaro D, Wopfer N, Huber CG, van Ree R, Mari A, Lackner P, Ferreira F, Schubert M, Gadermaier G. Hydrogen/deuterium exchange memory NMR reveals structural epitopes involved in IgE cross-reactivity

of allergenic lipid transfer proteins. J Biol Chem. 2020;295(51):17398–410. doi: 10.1074/jbc.RA120.014243. PubMed PMID: 33453986; PMCID: PMC7762950.

21. Wolbing F, Kunz J, Kempf WE, Grimmel C, Fischer J, Biedermann T. The clinical relevance of birch pollen profilin cross-reactivity in sensitized patients. Allergy. 2017;72(4):562–9. Epub 20160928. doi: 10.1111/all.13040. PubMed PMID: 27588729.
22. Asero R, et.al. Peanut allergy in Italy: a unique Italian perspective. J Allergy Clin Immunol: Global. 2022;1(2):61–6. doi: https://doi.org/10.1016/j.jacig.2022.02.001.
23. Dance A. The race to deliver the hypoallergenic cat. Nature. 2020;588(7836):S7–S9. doi: 10.1038/d41586-020-02779-3. PubMed PMID: 33268857.
24. Santos SPO, Lisboa ABP, Silva FSR, Tiwari S, Azevedo V, Cruz AA, Silva ES, Pinheiro CS, Alcantara-Neves NM, Pacheco LGC. Rationally designed hypoallergenic mutant variants of the house dust mite allergen Der p 21. Biochim Biophys Acta Gen Subj. 2022;1866(4):130096. Epub 20220122. doi: 10.1016/j.bbagen.2022.130096. PubMed PMID: 35077824.
25. Tscheppe A, Palmberger D, van Rijt L, Kalic T, Mayr V, Palladino C, Kitzmuller C, Hemmer W, Hafner C, Bublin M, van Ree R, Grabherr R, Radauer C, Breiteneder H. Development of a novel Ara h 2 hypoallergen with no IgE binding or anaphylactogenic activity. J Allergy Clin Immunol. 2020;145(1):229–38. Epub 20190913. doi: 10.1016/j.jaci.2019.08.036. PubMed PMID: 31525384; PMCID: PMC7100897.
26. Lewkowich IP, Day SB, Ledford JR, Zhou P, Dienger K, Wills-Karp M, Page K. Protease-activated receptor 2 activation of myeloid dendritic cells regulates allergic airway inflammation. Respir Res. 2011;12:122. Epub 2011/09/23. doi: 1465-9921-12-122 [pii]10.1186/1465-9921-12-122. PubMed PMID: 21936897; PMCID: 3184630.
27. Page K, Ledford JR, Zhou P, Dienger K, Wills-Karp M. Mucosal sensitization to German cockroach involves protease-activated receptor-2. Respir Res. 2010;11:62. Epub 2010/05/26. doi: 1465-9921-11-62 [pii] 10.1186/1465-9921-11-62. PubMed PMID: 20497568; PMCID: 2889872.
28. Porter P, Susarla SC, Polikepahad S, Qian Y, Hampton J, Kiss A, Vaidya S, Sur S, Ongeri V, Yang T, Delclos GL, Abramson S, Kheradmand F, Corry DB. Link between allergic asthma and airway mucosal infection suggested by proteinase-secreting household fungi. Mucosal Immunol. 2009;2(6):504–17. Epub 2009/08/28. doi: mi2009102 [pii] 10.1038/mi.2009.102. PubMed PMID: 19710638.
29. Wills-Karp M. Allergen-specific pattern recognition receptor pathways. Curr Opin Immunol. 2010;22(6):777–82. Epub 2010/11/26. doi: S0952-7915(10)00161-5 [pii] 10.1016/j.coi.2010.10.011. PubMed PMID: 21093238.
30. Le LQ, Mahler V, Scheurer S, Foetisch K, Braun Y, Weigand D, Enrique E, Lidholm J, Paulus KE, Sonnewald S, Vieths S, Sonnewald U. Yeast profilin complements profilin deficiency in transgenic tomato fruits and allows development of hypoallergenic tomato fruits. FASEB J. 2010;24(12):4939–47. Epub 20100813. doi: 10.1096/fj.10-163063. PubMed PMID: 20709910.

31. Neelakandan AK, Wright DA, Traore SM, Ma X, Subedi B, Veeramasu S, Spalding MH, He G. Application of CRISPR/Cas9 system for efficient gene editing in peanut. Plants (Basel). 2022;11(10). Epub 20220520. doi: 10.3390/plants11101361. PubMed PMID: 35631786.
32. Hilu KW, Friend SA, Vallanadu V, Brown AM, Hollingsworth LRt, Bevan DR. Molecular evolution of genes encoding allergen proteins in the peanuts genus Arachis: structural and functional implications. PLoS One. 2019;14(11):e0222440. Epub 20191101. doi: 10.1371/journal.pone.0222440. PubMed PMID: 31675366; PMCID: PMC6824556.

5 The Two Faces of Interferon

The Roman god, Janus, has two faces, looking into both the past and the future.

OVERVIEW

1. Cytokines, including IFNs and interleukins, are intercellular messengers that are usually produced in vanishingly small amounts in the body. They have very high specific activities (measured in millions of units/mg protein, where a unit is typically the smallest amount of a protein that can be measured in an activity assay).
2. IFNs were first identified in supernatants of stimulated white blood cells. Limited quantities prepared from the "buffy coat" fraction of human blood prevented severe virus infections.
3. IFN could be used clinically after cloning and recombinant expression allowed large quantities to be produced.
4. There are many different IFN proteins, of which certain IFN-α isoforms and IFN-β have had the most success in the clinic, for treating viral infections, multiple sclerosis (MS) and certain cancers.
5. The different effects of IFNs and other cytokines are exerted through binding to receptors on cells and inducing a "kinase cascade" that stimulates the production of other factors within the cell. Kinases add phosphate groups to proteins, usually to start the cascade, while phosphatases remove these to stop it.
6. Antiviral proteins like interferons (IFNs) and other cytokines involved in immune system regulation are removed from circulation within hours or days. These protein induce interferon-stimulated genes (ISGs) and suppressors that stop their synthesis (SOCS, SHPs) as part of a complex regulatory system.
7. Many viruses turn off IFN directly by degrading factors that stimulate their synthesis (using normal cell pathways called autophagy or necrosis) or through activating proteins such as SOCS that naturally regulate IFN and cytokine synthesis. SOCS inhibitors can enhance IFN's antiviral activity.

DOI: 10.1201/9781003333319-6

8. Although many lives have been improved or saved by IFN therapy for cancer and viral infections, long-term clinical use illustrates how these valuable proteins can be conditionally toxic.
9. Treatments for diseases caused by overproduction of IFN include monoclonal antibodies that prevent IFN binding to receptors on the cell surface or inhibiting the activity of kinases which initiates the ISG production cascade.

THE POSITIVE AND NEGATIVE SIDES OF CYTOKINES

Mutations that cause changes in activities of normal human proteins can lead to serious diseases. The next chapters deal with an even more startling fact: that natural signaling and control proteins produced by the human body, essential for eliminating pathogens, can also cause acute or chronic diseases. While antibodies are generally considered to be essential for the body's response to pathogens, the process of producing specific antibodies requires weeks to months, which means they can only fight off later stages of an infection, or subsequent infection with a similar pathogen. Antibodies may play an earlier role if previously induced by vaccines. Unless previously induced by vaccination, specific antibody protection against pathogens only begins to be effective after about a week, starting with large, relatively non-specific IgM molecules. Smaller, more specific IgG antibodies (which typically have higher affinity) first appear after two to three weeks. In response to some stimuli, B-cells can also produce IgE antibodies, which play a role in allergy (see the previous chapter).

But the first line of defense against a virus or other pathogen is mediated by our innate immune system, a system of proteins that detect and protect against invaders. This chapter introduces a vital part of that response, a family of small proteins, the interferons (IFNs), named as they are substances produced and secreted by blood cells that "interfere with" virus growth.[1]

VIRUS EXCLUSION AND THE DISCOVERY OF INTERFERON

This story stretches over many decades, to the earliest days of cell research. Researchers around the world were intrigued by "virus exclusion", where infection with one virus, or treating uninfected cells with the supernatant of infected cells, might prevent infection by another virus (Figure 5.1). However, the phenomenon was difficult to study in the first half of the 20th century for various reasons, including the difficulty of maintaining

FIGURE 5.1 IFNs were identified as a soluble factor ("s") that could induce cells to develop protection against viruses. IFN can be induced by a virus or adding double-stranded RNA (dsRNA) to the medium of cultured cells. The secreted IFN binds to a receptor on the surface of other cells, inducing them to produce proteins from IFN-stimulated genes (ISGs) that render them resistant to virus infection ("virus exclusion").

a sterile environment to grow mammalian cells. Further, the interfering activity was not easily demonstrated reproducibly. Too much live virus would kill the cells before they secreted a significant quantity into their growth medium; too little, and there was no response to infection. The secreted activity was also heat sensitive and disappeared on freezing and storage. IFNs are produced in tiny amounts by cells and exert their effects at picomolar concentrations. The activity in supernatants disappeared with heating, freezing or simple storage. The early researchers seemed to be chasing a *fata morgana*.

Despite all these difficulties, Alick Isaacs (a Scottish MD and a Rockefeller Research Fellow), working with Margaret Edney in Melbourne Australia, showed in 1950 that fertilized eggs, pretreated with heat-inactivated influenza virus, were less susceptible to infection by a live flu virus.[2] Their interpretation was that some substance, essential for the growth of the virus, was depleted in the treated eggs. Isaacs changed his mind after continuing the work in England. He and others, including Jean Lindenmann, a new medical graduate from Switzerland, established by 1957 that filtered supernatants of treated egg membranes could protect untreated membranes from virus.[3,4] The stimulated cells produced a specific substance, probably a protein, which interfered with the growth of four quite different viruses: human influenza (a segmented, negative-strand RNA virus), Newcastle Disease Virus of chickens (negative-strand, non-segmented RNA, a paramyxovirus in the same family as measles/mumps), Sendai (a negative-strand RNA, paramyyxovirus of mice) and vaccinia (a large DNA virus best known for inducing antibodies against smallpox). The experiments convinced the world that IFN was real, but it was still mysterious and undefined.

COUNTING THE CHICKENS

For the next 20 years, researchers from industry and universities tried to isolate IFN from fertilized chicken eggs or induced human white blood cells. Russian groups reported success early and treated hundreds of people during the influenza epidemic of 1969 with an IFN preparation. As their research reports were published almost exclusively in Russian journals, the rest of the world was not convinced. The Soviet scientists involved were also puzzled by the results of treatment. They concluded that whether a person responded or not was probably due to their own intrinsic ability to make IFN (since treating cells with IFN prompts them to make more IFN). The over-the-counter IFN preparations sold in East block pharmacies for treating flu could possibly contain an agent used to induce IFN (such as viral fragments, or chemically synthesized dsRNA or poly I/C). They did however convince the world that it was possible to produce an antiviral preparation for use in humans.

Efforts to isolate IFN in quantities sufficient to treat patients continued, although many groups gave up after realizing the difficulty of the task. For example, one group at a Swiss company worked for ten years, using hundreds of thousands of virus-treated eggs. They obtained preparations that contained several million IFN units (IFU) per mg of protein, removing 99.9% of the inactive contaminants in the starting egg extract. However, even this preparation (about 10,000× cleaner than what they started with) was a complex mixture of different proteins, with none individually associated with antiviral activity. Further, chick IFN could not be used to treat humans. Besides the anticipated problems with immune reactions to contaminating proteins, IFNs are species specific. Seeing a long and expensive battle ahead to obtain a therapeutic agent, the company halted the project in the early 1970s and the group dispersed.

FINNISH BLOOD BANK LEADS THE WAY

Human IFN for clinical testing was produced in meaningful amounts by the group of Kari Cantell, at the Finnish National Health Center. They purified IFN from the "buffy coat" fraction, a brownish layer (making less than 1% of total blood volume), which lies between the plasma and the red cell layer when whole blood is centrifuged (Figure 5.2, top). Cantell wrote that the mixture of leucocytes and granulocytes in the

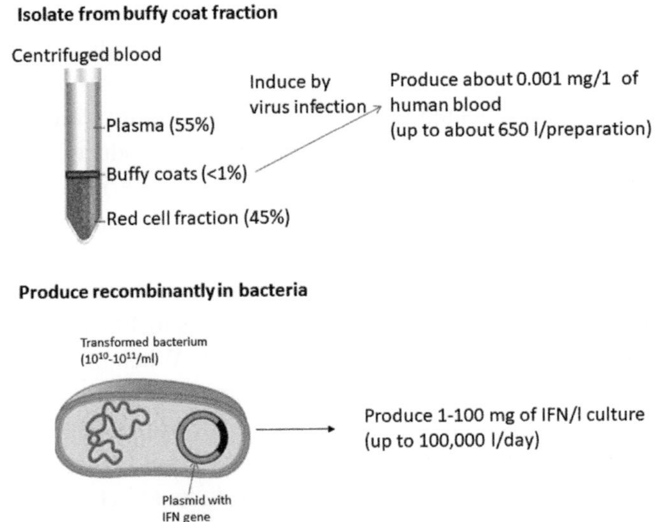

FIGURE 5.2 Cloning of IFN permitted essentially unlimited production. Recombinant protein can be made much more efficiently than that isolated from human leukocytes (blood cells). It is easier to purify and use for therapy.

buffy coats, leftovers from the plasma and packed red blood cell fractions sent to the clinic, would otherwise go to waste, as "no one else could find a use for them".

Thanks to support from the Blood center of the Finnish Red Cross, the Cantell group was able to produce IFN from leukocytes in 65-liter volumes, each representing more than 300 liters of initial human blood. From 500 ml blood, they could obtain 12 ml containing about 10^9 packed buffy coat cells, which they distributed to about 100 ml of culture medium (10^8 cells/ml) to which they added virus as an inducing agent for the IFN activity. After several days, they purified the supernatant of the cultures by rounds of precipitation with different agents, to obtain protein mixtures with 3–5 million IFU/mg protein.[5]

Using all the available human blood supplies in Finland, they were able to produce several billion ($3–5 \times 10^9$) IFU* total. This small group became, for a time, the largest distributor of human IFN (still only 1–3% pure) in the

* IFU is interferon units. The specific activity of IFNs range from 10^7 (recombinant IFN-γ) to 2×10^8 (recombinant IFN-$\alpha 2$) IFU/mg protein, depending on the cells and virus used for the assays. The total amount of IFN produced from the blood bank was on the order of 100 mg, or about enough to treat 200–500 patients for a short time. While the other cell proteins in the early extracts were ignored, we now know that even ng levels of contaminants can cause immune reactions upon repeated injection of insulin, cytokine and antibody preparations.

world. The precious preparations were distributed to clinical groups and used with some success to treat diseases ranging from fulminating hepatitis to cancers. However, the methods were labor and blood intensive. The first studies showed repeated doses of 1–5 million units were needed per patient. From 2000 blood donations, Cantell's group produced about 20 mg of IFN, or only enough to treat 200 patients for short periods of time.

Fortunately, just as the early studies validated the antiviral activities of IFN, the gene for the first IFN was finally identified, following closely after the cloning of insulin.[6] IFN-α2 was cloned in the group of Charles Weissmann at the University of Zürich (researchers included Shiga Nagata, Hideharu Taira, Michel Streuli and many others).[7] When inserted into bacterial cells, which are easier to grow than human cells, it was possible to make much more protein. Recombinant IFN-α2 (i.e., from the cloned gene, placed into a bacterial expression system, lower part of Figure 5.2) allowed, for the first time, clinical studies to demonstrate its effects in humans. By directing bacteria to turn out mgs of the protein per liter of culture, more IFN could be produced in a single industrial scale fermenter run than from all the cell cultures of all the groups in the world in the decades after the findings of Isaac's group.[†] The purified recombinant protein proved as effective an antiviral as the IFN preparation made from blood cells.[8]

Thanks to the fact that there were few treatments at that time for viral infections or cancer, the cloning of IFN was widely publicized, even making the cover of Time magazine!

IFNs induce an army of messengers against pathogens. Soon after infection, IFN signals other cells to make ISGs (interferon stimulated genes) to protect themselves. As Figure 5.3 summarizes, IFN begins to rise as soon as the first virus particles are detected by cells.

IFN binds to specific cell receptors, activating "JAK/STAT" kinases, which in turn stimulate the production of ISGs, leading to production of proteins that help them resist any virus (virus exclusion) and many other pathogens. Synthesis of IFN itself is rapidly turned off as virus levels fall, along with other cytokines, by a group of "suppressors of cytokine signaling" (SOCS). Specific phosphatases (SHPs) can also stop the cascade leading to ISGs. Many viruses activate SOCS or directly degrade ISGs to inhibit IFN's effects. Conversely, a small peptide that inhibits SOCS enhances antiviral activity.

IFN needs to be removed from the blood stream to enable cells to resume growth. Other intercellular messengers or cytokines (a group that

---

† Alick Isaacs did not live to see IFN's validation. He died from intracranial hemorrhage after several years of illness in 1967 at the age of 45. His Swiss colleague, Jean Lindenmann, continued studying the IFN system until about 2007 and died in 2015 at the age of 90.

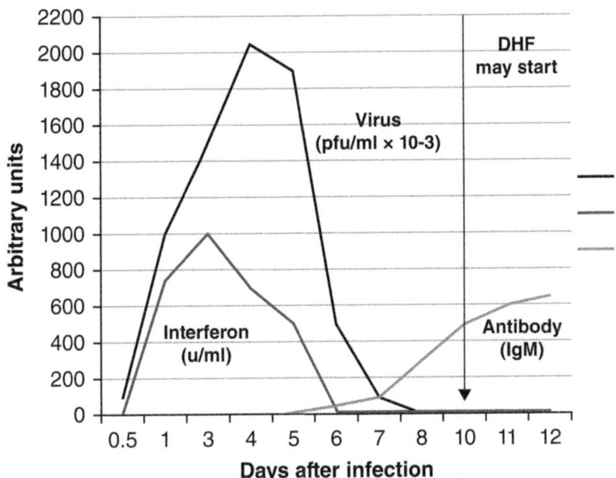

FIGURE 5.3 IFN levels in serum rise within hours after virus infection from nearly undetectable to about 800 IFU/ml. The virus titer continues to rise for a few days (to 10^7–10^9 plaque forming units/ml) but drops as the first (less specific) IgM antibody levels begin to rise. The first specific IgG antibodies develop weeks later, often after there is no measurable virus in the blood.

includes interleukins, chemokines as well as diverse factors still named for their observed activities, such as leukemia inhibitory protein and tumor necrosis factor (TNF) may have longer half-lives. The role of IFNs and ISGs' protein products as the first line of defense against infection cannot be ignored: many patients suffering from viral pneumonias, including many unvaccinated patients with severe Covid-19, had mutations that reduced response to IFNs.[9] A large proportion (at least 10%, more depending on study) of severely ill Covid-19 patients (of all ages) made antibodies (autoantibodies) blocking their own IFNs.[10]

IFNs WERE AMONG THE FIRST PHARMACEUTICALLY USEFUL PROTEINS

Subsequent cloning of IFN protein genes (Figure 5.4) revealed that the active components were a mixture of many different related proteins.[7] Cloning also revealed that although there are many Type 1 IFN-αs, there is only one IFN-β gene. Because Type 1 IFNs all have some sequence similarity, they were cloned relatively rapidly. Several years went by before IFN-γ (Type 2) was cloned (Figure 5.4). Type 2 and 3 IFNs proved to have different sequences from any Type 1 IFN and used different receptors. Although IFN-γ (Type 2) folded into a similar 3-D structure (Figure 5.5), the IFN-γ dimer showed the helices entwined with one another, instead of the more

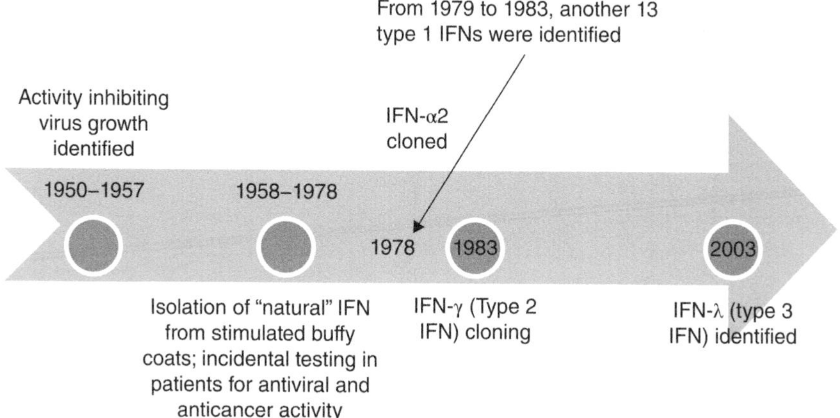

FIGURE 5.4 Cloning revealed that there are 14 related Type 1 IFNs that bind a common receptor. Type 2 (IFN-γ) and Type 3 (IFN-λ) have different sequences, receptors and activities.

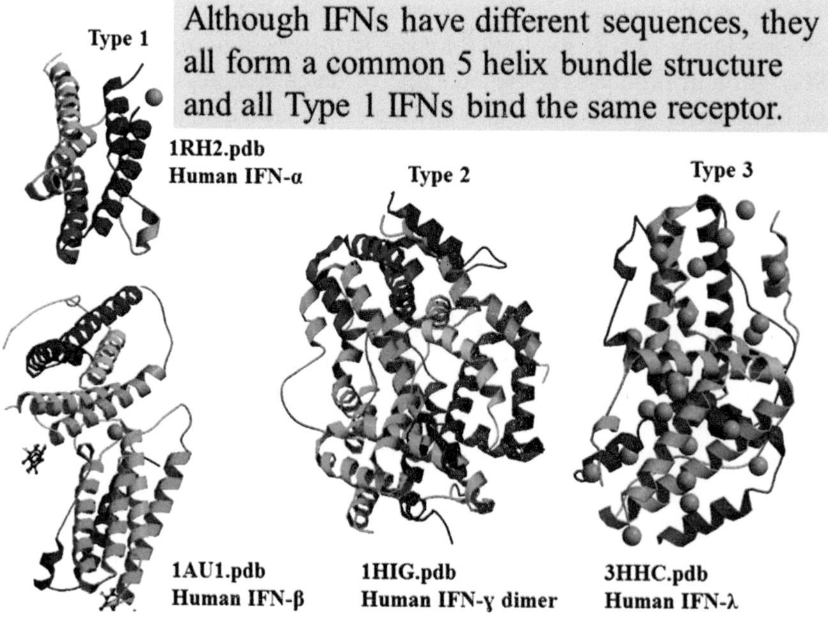

FIGURE 5.5 IFN molecular structures, taken from files in the protein database (PDB), show a similar helical bundle, called a "cytokine fold".

open dimeric structure of IFN-β. About 20 years later, yet a third type of IFN, IFN-λs (Type 3) were identified. Although IFN-λ sequences had similarity to Type 1 IFNs, they had a distinct receptor and discrete activities of their own. In general, Type 1 and 2 IFNs are active throughout the body, while the receptors for IFN-λ are on epithelial cells. While Type 1 IFNs can protect against Influenza virus if administered systemically, Type 3 IFN protects if infection of the virus is intranasal (i.e., inhaled through the nose).[11]

The fact that the body makes so many different IFNs shows how essential their activities are. However, deleting one IFN gene is unlikely to be lethal as they all have similar effects on cells. Since IFN α's and β (Type 1 IFNs) all use the same receptor on cells, mutations in the receptor (or blocking it, as we will see later) can lead to much more immune suppression than deleting one IFN protein. Why all these related proteins with a central activation center are needed became clearer as other functions of IFNs were revealed. They may have different abilities to induce proteins such as ribonucleases[12,13] or specific ISGs.[14]

The path to the clinic for recombinant IFNs (Figure 5.6) was relatively rapid thanks to the early success of the leukocyte IFN preparations. Although the active protein percent of leukocyte preparations was low, they treated virus infections (including fulminating hepatitis in pregnant women), chronic tumors and even sarcomas (hence the promise of IFN as an anticancer drug). Many individuals, convinced that IFN would cure all cancers, clamored for IFN treatment, even before the recombinant protein was available or approved for use. After showing that the bacterially produced IFN had the same biological activities as the leucocyte preparations,[8] recombinant IFN was approved to treat a blood cell cancer, hairy cell leukemia. Figure 5.6 illustrates how quickly approval was obtained for treating chronic viral infections and various forms of cancer. By the 1990s,

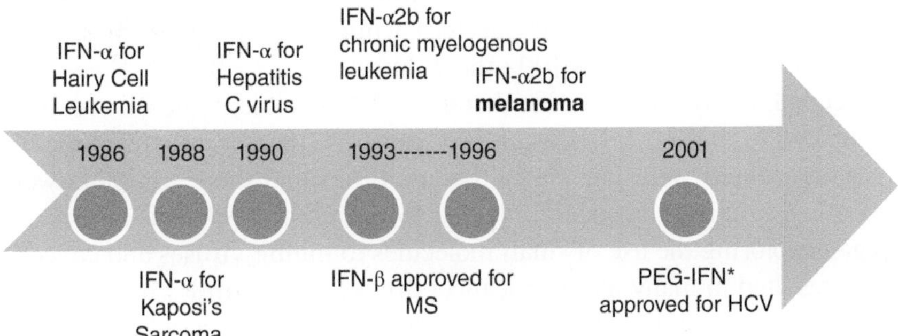

FIGURE 5.6 Recombinant interferons quickly found their way into the clinic and received FDA approval for treating various diseases.

IFN-α had been approved for treating papillomavirus and herpes viruses. In general, treatment of chronic viral diseases was more successful than for acute infections. The cost of treatment with IFNs was too high to consider treating common viruses from which the patient could be expected to recover with few side effects.

Today, IFN-α preparations are used to treat hepatitis C and papilloma virus, as a sole or combination therapy for a variety of cancers. While IFN-αs were not effective for treatment of multiple sclerosis (MS), IFN-β gave significant improvement in symptoms and became part of advanced treatments. Attaching polyethylene glycol, a long fatty acid like molecule, can extend the half-life of IFNs in the blood, meaning less frequent injections. Several different PEG-IFN preparations have been approved for clinical use in MS, meaning fewer injections for patients.

PROBLEMS WITH TREATING WITH IFN PROTEIN INFUSIONS

The timeline in Figure 5.6 slows around 2001 for several reasons. First, although many diseases continue to be treated with IFNs, and the FDA continues to approve new uses, they are being superseded in many areas. Side effects, the need for injection and costs of IFN treatment have led to the use of more recently introduced specific antibodies directed against B-cells and oral medications for MS therapy. Side effects were probably inevitable, with treatment raising levels of IFN to such a degree for a long period of time. IFN used to treat chronic hepatitis C virus infections or MS was now a permanent resident, instead of occurring occasionally in the sera of patients. Also, many early studies used too much IFN, as the specific activity of the recombinant proteins was so much higher than the impure preparations made from buffy coats that were tested initially. Current dosages are in the 100,000–8 million unit/day range (i.e., usually about 1–10 μg of the protein is injected), while early *in vitro* studies would use (and waste) as much as 100× those doses.

Second, many more specific therapies are available in the 21st century, compared to the late 1970s when IFN was first brought into the clinic. At that time, there were few chemotherapy drugs and those available were toxic and relatively unspecific. In the 1990s, companies and universities began exploring the use of small molecules to inhibit viruses and cancers. This resulted in many more anticancer drugs and antivirals.

Third, antibody therapies against specific proteins that, when mutated, could drive cancers (against, for example, the epidermal growth factor receptor, HER2, which stimulates the growth of about 20% of breast cancers) were also developed, piggybacking on the improved methods for producing recombinant proteins that had been introduced for IFN, insulin, clotting factors and other cytokines. Also, as our understanding of IFNs' effects deepened, it became possible to design better therapies. For example, understanding the effects of IFN-β that improved MS symptoms allowed the design of small molecules and antibodies specifically targeting pathways leading to myelin degradation.

THE JANUS FACE OF IFNs IS REVEALED

Finally, while IFN treatments are considered relatively non-toxic, there were significant side effects when they were used over long periods of time. Early reports of potential IFN toxicity[15,16] were attributed to the contaminants in the preparations. However, major side effects (flu-like symptoms, fatigue, nausea/anorexia, neutropenia, neuropsychiatric symptoms [depression] and injection site reactions) could not be ignored in patients treated with pure preparations of the recombinant proteins. Treating a cold with IFN generated symptoms resembling the infection itself.

LIKE OTHER VITAL CELLULAR REGULATORS, IFNs HAVE INTRINSIC TOXICITY

Insulin is the best known of the essential proteins whose levels must be strictly regulated to avoid toxic effects. All cells require insulin for growth, but direct injection can lead to insulin overdosing. This can be deadly in diabetics if not countered with rapid infusion of glucose. Insulin has also been used for precisely these (presumably rapidly reversible) toxic effects, in insulin shock therapy used in psychiatry.

Along with clinical success came recognition of IFNs' potential toxicity and the role naturally overproduced or injected recombinant IFNs may play in exacerbating autoimmune diseases, such as systemic lupus erythematosus (SLE). Recognizing the autoimmune effects of extended IFN treatment also revealed a link to this chronic disease.

Lupus and IFN: SLE is an autoimmune disease characterized by rash and swollen lymph nodes,[17] where patients become incapacitated with severe nephritis and eventual kidney failure. Unexpectedly, some of the patients treated with Type 1 IFN for MS developed lupus symptoms, which could resolve when treatment was terminated. Further studies showed that some patients with SLE, who had never received IFN therapy, had high circulating levels of IFN and ISGs. Thus, treating patients with IFN suggested a new way to treat SLE.

SLE today is considered to be at least partially an IFN-induced disease, with about 35% of patients, especially those showing the highest "signatures" of IFN and certain ISGs in their blood, responding to treatment with inhibitors of IFN. The current theory is that those with a genetic predisposition will develop SLE after some trigger, which can include IFN treatment of an underlying disease (e.g. MS), or a severe viral infection. Thus, while IFN continues to be used for many indications, there is now more caution in prescribing it. One should use as little as possible and limit treatment times.

Several therapies designed to neutralize IFN have reached FDA approval for treating SLE[18] (Figure 5.7), although none completely eliminates the disease or is effective in all patients. The first antibodies, designed to bind to IFNs directly, had disappointing results. Rontalizumab,[19] an antibody specific for IFN-α2, showed only some benefit in patients with a low IFN gene score,[20] suggesting its affinity for the cytokine was too weak. Another antibody against IFN, sifalimumab, reached Phase II trials[21] before being abandoned for a more successful one, Anifrolumab.

Anifrolumab binds to the common IFN receptor, thus inhibiting multiple Type 1 IFNs. Two international "phase 3" studies, called TULIP 1 and 2, indicated that anifrolumab could lower SLE score and reduce steroid use. About 35% of SLE patients treated with anifrolumab had an improved disease score, compared to 18% of placebo.[22] The results showed a greater effect in patients shown to be producing more IFN at the start of treatment.

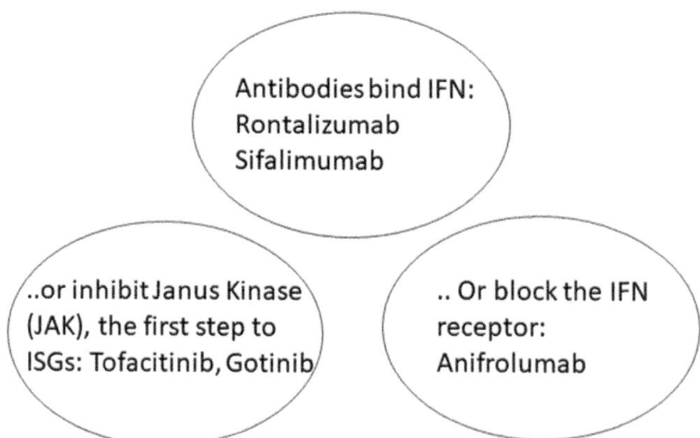

FIGURE 5.7 **There are many ways to inhibit the effects of interferons,** for example, by specific antibodies that bind to IFNs directly or their receptors. Antibodies to the common type 1 IFN receptor have had the most success in treating autoimmune diseases such as SLE.

Once again, the results were mixed, because blocking IFN to control autoimmunity also blocks its antiviral effects. Herpes virus infections increased in the treated group, indicating that their natural IFN levels were indeed instrumental in controlling this virus.[23] Fortunately, a more extended trial indicated that patients treated with anifrolumab for three years did not have higher rates of serious infections compared to placebo, or if vaccinated, of Covid-19.[24] Time will tell whether anifrolumab will prove sufficiently better at symptom control than less expensive therapies for SLE patients, such as prednisone or other corticosteroids, mizoribine, mycophenolate mofetil or mycophenolic acid, methotrexate or the antimalarial, azathioprine.

Janus kinase (JAK) and other inhibitors. There are many other antibodies and even small molecules being tested to control IFN levels. For example, another approach is to target past the level of IFN binding to the parts of its receptor. After IFN binds to its receptor, receptor associated proteins, called JAK, become phosphorylated (i.e., a phosphate group is added to a tyrosine residue) and activated. Several orally available "jakinibs", small molecule inhibitors of JAKs are approved or in testing (Tofacitinib, Gotinib etc.). The structures of all these drugs have some common elements, but a lot of diversity (Figure 5.8). They work to prevent phosphate groups from being added to tyrosine residues at the end of the IFN receptor. Combining IFN directed therapeutics with inhibitors of other proteins may also help in autoimmune diseases. For example, elevated levels of B-lymphocyte stimulator (BLyS), required for survival and maturation of B-cells to produce antibodies, are also found in autoimmune diseases. The jakinib Benlysta (Belimumab) specifically recognizes and binds to BLyS, inhibits BLyS's stimulation of B-cell development and, finally, restores the

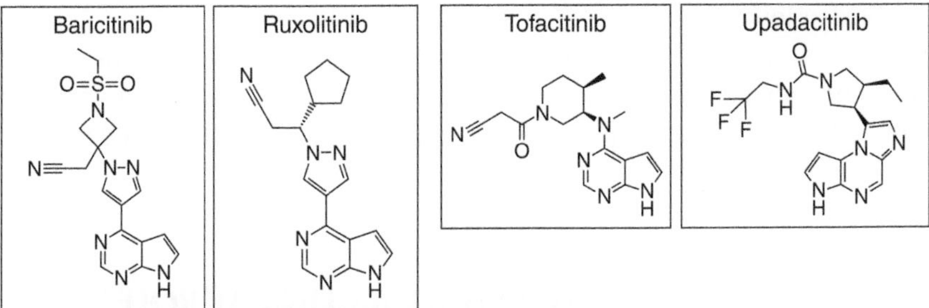

FIGURE 5.8 Small molecules can be used to inhibit the JAK-STAT kinases that are activated by phosphorylation when IFNs bind to cells. Kinase inhibitors also can be used to inhibit the effects of other cytokines, as will be seen in the next chapter.

potential for autoantibody-producing B-cells to undergo the normal process of apoptosis (programmed cell death).

Jakinibs are also used for treating other autoimmune diseases. Their function can depend on binding to other cell receptor kinases besides IFN's. One successful SLE drug, baricitinib (Olumiant), a JAK 1&2 inhibitor, was approved to treat rheumatoid arthritis (RA) in 2018. Trials of baricitinib in SLE and eczema are ongoing. Upadacitinib, a Jak1 inhibitor, induced clinical remission in about half of RA patients, versus just 18% among those given a standard first-line treatment. It is also used to treat other autoimmune diseases, including IBD and Crohn's disease.

Kinase inhibitors may also help in rare disorders, including myeloproliferative neoplasms such as polycythemia vera. A JAK2 inhibitor, ruxolitinib (Incyte's *Jakafi*) can replace bloodletting in patients resistant to hydroxyurea who develop an enlarged spleen. Further trials are thus needed to determine which inhibitors and combinations are best for RA, SLE or other autoimmune syndromes.

JAKINIBS HAVE THEIR OWN "OFF-TARGET" ACTIVITIES LEADING TO SIDE EFFECTS

As noted above, some Jakinibs also inhibit other tyrosine kinases involved in pathways important for cell survival. As with antibodies that directly inhibit IFN, they greatly reduce the body' ability to deal with chronic viral infections. Use of Jakinibs in RA and IBD patients has now been shown to be related to development of herpes zoster virus rashes and pain (shingles), perhaps by reactivation of a dormant virus. It is thus recommended that patients receive inactivated Shingrix vaccine before starting the treatment. Patients' resistance to other infections, such as tuberculosis, is also lowered.

Other serious side effects, including cardiovascular problems, blood clots and cancers, were found in a long-term study of RA patients treated with jakinibs. These "phase 4" (post-approval) clinical trials led to a "black box warning" from the FDA recommending caution in their use, especially in older patients with additional risk factors.[25]

INHIBITORS OF IFN-γ ALSO HAVE A CLINICAL ROLE

While Type 1 IFN was quickly taken up in the clinic, IFN-γ proved more difficult to use. One reason for this is that this powerful cytokine has so many different activities, such as slowing cell growth, that are hard to separate from its' positive anti-pathogen effects. Induced in many disease states and important for the immune response to many pathogens, including

Covid-19, the "acid sensitive" IFN-γ may yet have a clinical future. More recent studies have indicated elderly patients produce lower levels of IFN-γ in response to vaccination than younger ones.[26] Elderly individuals survive pneumonia better if they produce more IFN-γ.

IFN-γ differs in many ways from the Type 1 or 3 IFNs. There is only one gene in humans for "Type 2" IFN, which also contains introns (these interruptions in the coding sequence of the gene made it particularly hard to identify using the reverse DNA sequence of the protein). Although the structures share some similarities, including the cytokine helical bundle[27,28] (see Figure 5.5), the sequence of IFN-γ is completely different. Further, it has its own receptor, the activation of which leads to distinct antiviral and growth inhibiting effects on cells. Its activities are also quite different with respect to its ability to activate the genes and activities of ribonucleases.[12,13,29]

While early studies did not show sufficient positive effects of treatment, blocking IFN-γ's activities proved more useful. For example, a soluble form of the IFN-γ receptor, which could block and prevent secreted IFN-γ from binding to its cell-bound receptor, prevented glomerulonephritis in a murine SLE model.[30] Much later, human studies with a monoclonal antibody against IFN-γ, AMG811, showed some preliminary promise,[31] but the treatment did not survive the many clinical trials required for demonstrating efficacy. More recent inhibitors of IFN-γ may eventually prove to be useful therapeutics in autoimmune diseases such as hemophagocytic lymphohistiocytosis (HLH), a rare and potentially life-threatening histiocytic disorder. Antigens stimulate an immune response accompanied by life-threatening overproduction of IFN-γ with inflammatory cytokines IL-10 and IL-6. A monoclonal antibody treatment, Emapalumab, that binds and neutralizes IFN-γ[32] was approved by the FDA in 2018 for primary HLH in adults and children with refractory, recurrent or progressive disease or who are intolerant to conventional HLH therapy.

CONCLUSION

The IFNs are the first line of defense against viruses, tuberculosis and many other pathogens. Cloning of human IFN genes allowed them to be produced in large amounts. They have been successfully used as antiviral drugs and against certain cancers. Pegylation can give longer half-lives in sera, meaning fewer injections, but the cost of therapy is still high. To date, protein therapeutics include three classes of IFN (α,β,γ; IFN-λ is still in clinical trials [e.g. NCT04354259] for its antiviral potential). However, long-term use of IFNs in the clinic, such as for treating MS, has been associated with depression and autoimmune diseases. A variety of novel therapies to reduce levels of IFN have yielded potential new therapies for SLE and other autoimmune syndromes.

The next chapter gives further evidence that while cytokines can have pharmaceutical uses, inhibitors of the same protein may also be treatments. Other cytokines, Interleukins 2, 7,12,15,21 (IL-2…IL-21) and tumor necrosis factor (TNF) have undergone extensive clinical testing for antiviral, cancer and specific disease therapy. But as with IFNs, the use of cytokines as protein therapeutics has many intrinsic problems, including cost, need for injection and often unanticipated side effects due to their many-sided activities. The next chapter highlights two cytokines with potential use in cancer therapy, if one can control their toxic activities on normal cells.

REFERENCES

1. Burke DC. Early days with interferon. J Interferon Cytokine Res. 2007; 27(2):91–6. doi: http://dx.doi.org/10.1089/jir.2007.9998. PubMed PMID: 215528533; 17316135.
2. Edney M, Isaacs A. Interference between inactive and active influenza viruses in the chick embryo. III. Inhibitor of virus haemagglutination in the chorioallantoic membrane. Aust J Exp Biol Med Sci. 1950;28(6):603–12. Epub 1950/11/01. PubMed PMID: 14838784.
3. Isaacs A, Lindenmann J, Valentine RC. Virus interference. II. Some properties of interferon. Proc R Soc Lond B Biol Sci. 1957;147(927):268–73. Epub 1957/09/12. PubMed PMID: 13465721.
4. Isaacs A, Lindenmann J. Virus interference. I. The interferon. Proc R Soc Lond B Biol Sci. 1957;147(927):258–67. Epub 1957/09/12. PubMed PMID: 13465720.
5. Ahstrom L, Dohlwitz A, Strander H, Carlstrom G, Cantell K. Letter: Interferon in acute leukaemia in children. Lancet. 1974;1(7849):166–7. Epub 1974/02/02. PubMed PMID: 4129735.
6. Goeddel DV, Kleid DG, Bolivar F, Heyneker HL, Yansura DG, Crea R, Hirose T, Kraszewski A, Itakura K, Riggs AD. Expression in Escherichia coli of chemically synthesized genes for human insulin. Proc Natl Acad Sci U S A. 1979;76(1):106–10. doi: 10.1073/pnas.76.1.106. PubMed PMID: 85300; PMCID: PMC382885.
7. Weissmann C, Nagata S, Boll W, Fountoulakis M, Fujisawa A, Fujisawa J-I, Haynes J, Henco K, Mantei N, Ragg H, Schein CH, Schmid J, Shaw G, Streuli M, Taira H, Todokoro K, Weidle U. Structure and expression of human alpha-interferon genes. Interferons: Elsevier; 1982. p. 295–326.
8. Schellekens H, de Reus A, Bolhuis R, Fountoulakis M, Schein CH, Ecsodi J, Nagata S, Weissmann C. Comparative antiviral efficiency of leukocyte and bacterially produced human alpha-interferon in rhesus monkeys. Nature. 1981;292(5825):775–6. doi: 10.1038/292775a0. PubMed PMID: 6167863.

9. Casanova JL, Abel L. From rare disorders of immunity to common determinants of infection: Following the mechanistic thread. Cell. 2022;185(17):3086–103. doi: 10.1016/j.cell.2022.07.004. PubMed PMID: 35985287; PMCID: PMC9386946.
10. Manry J, Bastard P, et al. The risk of COVID-19 death is much greater and age dependent with type I IFN autoantibodies. Proc Natl Acad Sci U S A. 2022;119(21):e2200413119. Epub 20220516. doi: 10.1073/pnas.2200413119. PubMed PMID: 35576468; PMCID: PMC9173764.
11. Gad HH, Dellgren C, Hamming OJ, Vends S, Paludan SR, Hartmann R. Interferon-lambda is functionally an interferon but structurally related to the interleukin-10 family. J Biol Chem. 2009;284(31):20869–75. Epub 20090520. doi: 10.1074/jbc.M109.002923. PubMed PMID: 19457860; PMCID: PMC2742852.
12. Schein CH, Haugg M. Deletions at the C-terminus of interferon gamma reduce RNA binding and activation of double-stranded-RNA cleavage by bovine seminal ribonuclease. Biochem J. 1995;307 (Pt 1):123–7. doi: 10.1042/bj3070123. PubMed PMID: 7717966; PMCID: PMC1136753.
13. Schein CH, Haugg M, Benner SA. Interferon-gamma activates the cleavage of double-stranded RNA by bovine seminal ribonuclease. FEBS Lett. 1990;270(1–2):229–32. doi: 10.1016/0014-5793(90)81275-s. PubMed PMID: 2121524.
14. Arimoto KI, Miyauchi S, Troutman TD, Zhang Y, Liu M, Stoner SA, Davis AG, Fan JB, Huang YJ, Yan M, Glass CK, Zhang DE. Expansion of interferon inducible gene pool via USP18 inhibition promotes cancer cell pyroptosis. Nat Commun. 2023;14(1):251. Epub 20230117. doi: 10.1038/s41467-022-35348-5. PubMed PMID: 36646704; PMCID: PMC9842760.
15. Farkkila M, Iivanainen M, Roine R, Bergstrom L, Laaksonen R, Niemi ML, Cantell K. Neurotoxic and other side effects of high-dose interferon in amyotrophic lateral sclerosis. Acta Neurol Scand. 1984;70(1):42–6. Epub 1984/07/01. PubMed PMID: 6206681.
16. Scott GM, Secher DS, Flowers D, Bate J, Cantell K, Tyrrell DA. Toxicity of interferon. Br Med J (Clin Res Ed). 1981;282(6273):1345–8. Epub 1981/04/25. PubMed PMID: 6165428; PMCID: 1504978.
17. Axtell RC, Raman C, Steinman L. Interferon-beta exacerbates Th17-mediated inflammatory disease. Trends Immunol. 2011;32(6):272–7. Epub 2011/05/03. doi: S1471-4906(11)00051-2 [pii] 10.1016/j.it.2011.03.008. PubMed PMID: 21530402; PMCID: 5414634.
18. Mok CC. Biological and targeted therapies of systemic lupus erythematosus: evidence and the state of the art. Expert Rev Clin Immunol. 2017:1–16. Epub 2017/04/27. doi: 10.1080/1744666X.2017.1323635. PubMed PMID: 28443384.
19. Maurer B, Bosanac I, Shia S, Kwong M, Corpuz R, Vandlen R, Schmidt K, Eigenbrot C. Structural basis of the broadly neutralizing anti-interferon-alpha antibody rontalizumab. Protein Sci. 2015;24(9):1440–50. Epub 2015/06/24. doi: 10.1002/pro.2729. PubMed PMID: 26099203; PMCID: 4570538.

20. Kalunian KC, Merrill JT, Maciuca R, McBride JM, Townsend MJ, Wei X, Davis JC, Jr., Kennedy WP. A Phase II study of the efficacy and safety of rontalizumab (rhuMAb interferon-alpha) in patients with systemic lupus erythematosus (ROSE). Ann Rheum Dis. 2016;75(1):196–202. Epub 2015/06/04. doi: annrheumdis-2014-206090 [pii] 10.1136/annrheumdis-2014-206090. PubMed PMID: 26038091.
21. Khamashta M, Merrill JT, Werth VP, Furie R, Kalunian K, Illei GG, Drappa J, Wang L, Greth W. Sifalimumab, an anti-interferon-alpha monoclonal antibody, in moderate to severe systemic lupus erythematosus: a randomised, double-blind, placebo-controlled study. Ann Rheum Dis. 2016;75(11):1909–16. Epub 2016/03/25. doi: annrheumdis-2015-208562 [pii] 10.1136/annrheumdis-2015-208562. PubMed PMID: 27009916; PMCID: 5099191.
22. Furie R, Khamashta M, Merrill JT, Werth VP, Kalunian K, Brohawn P, Illei GG, Drappa J, Wang L, Yoo S. Anifrolumab, an anti-interferon-alpha receptor monoclonal antibody, in moderate-to-severe systemic lupus erythematosus. Arthritis Rheumatol. 2017;69(2):376–86. Epub 2017/01/29. doi: 10.1002/art.39962. PubMed PMID: 28130918; PMCID: 5299497.
23. Morand EF, Furie R, Tanaka Y, Bruce IN, Askanase AD, Richez C, Bae SC, Brohawn PZ, Pineda L, Berglind A, Tummala R, Investigators T-T. Trial of anifrolumab in active systemic lupus erythematosus. N Engl J Med. 2020;382(3):211–21. Epub 20191218. doi: 10.1056/NEJMoa1912196. PubMed PMID: 31851795.
24. Kalunian KC, Furie R, Morand EF, Bruce IN, Manzi S, Tanaka Y, Winthrop K, Hupka I, Zhang LJ, Werther S, Abreu G, Hultquist M, Tummala R, Lindholm C, Al-Mossawi H. A randomized, placebo-controlled phase III extension trial of the long-term safety and tolerability of anifrolumab in active systemic lupus erythematosus. Arthritis Rheumatol. 2022. Epub 20221111. doi: 10.1002/art.42392. PubMed PMID: 36369793.
25. Winthrop KL, Cohen SB. Oral surveillance and JAK inhibitor safety: the theory of relativity. Nat Rev Rheumatol. 2022;18(5):301–4. Epub 20220322. doi: 10.1038/s41584-022-00767-7. PubMed PMID: 35318462; PMCID: PMC8939241.
26. McElhaney JE, Verschoor CP, Haynes L, Pawelec G, Loeb M, Andrew MK, Kuchel GA. Key determinants of cell-mediated immune responses: a randomized trial of high dose vs. standard dose split-virus influenza vaccine in older adults. Front Aging. 2021;2. Epub 20210521. doi: 10.3389/fragi.2021.649110. PubMed PMID: 35128529; PMCID: PMC8813165.
27. Schein CH. The shape of the messenger: using protein structure information to design novel cytokine-based therapeutics. Curr Pharm Des. 2002;8(24):2113–29. doi: 10.2174/1381612023393161. PubMed PMID: 12369857.
28. Schein CH. From interleukin families to glycans: relating cytokine structure to function. Curr Pharm Des. 2004;10(31):3853–5. doi: 10.2174/1381612043382512. PubMed PMID: 16381102.
29. Schein CH. From housekeeper to microsurgeon: the diagnostic and therapeutic potential of ribonucleases. Nat Biotechnol. 1997;15(6):529–36. doi: 10.1038/nbt0697-529. PubMed PMID: 9181574.

30. Ozmen L, Roman D, Fountoulakis M, Schmid G, Ryffel B, Garotta G. Experimental therapy of systemic lupus erythematosus: the treatment of NZB/W mice with mouse soluble interferon-gamma receptor inhibits the onset of glomerulonephritis. Eur J Immunol. 1995;25(1):6–12. doi: 10.1002/eji.1830250103. PubMed PMID: 7843255.
31. Welcher AA, Boedigheimer M, Kivitz AJ, Amoura Z, Buyon J, Rudinskaya A, Latinis K, Chiu K, Oliner KS, Damore MA, Arnold GE, Sohn W, Chirmule N, Goyal L, Banfield C, Chung JB. Blockade of interferon-gamma normalizes interferon-regulated gene expression and serum CXCL10 levels in patients with systemic lupus erythematosus. Arthritis Rheumatol. 2015;67(10):2713–22. doi: 10.1002/art.39248. PubMed PMID: 26138472; PMCID: PMC5054935.
32. Locatelli F, Jordan MB, Allen C, Cesaro S, Rizzari C, Rao A, Degar B, Garrington TP, Sevilla J, Putti MC, Fagioli F, Ahlmann M, Dapena Diaz JL, Henry M, De Benedetti F, Grom A, Lapeyre G, Jacqmin P, Ballabio M, de Min C. Emapalumab in children with primary hemophagocytic lymphohistiocytosis. N Engl J Med. 2020;382(19):1811–22. doi: 10.1056/NEJMoa1911326. PubMed PMID: 32374962.

6 Storming Cytokines in Infections and Cancers

… in a considerable number of cases of inoperable cancer of all varieties, and especially sarcoma, such tumours have been known entirely to disappear under attacks of accidental erysipelas, and patients have remained well for many years thereafter.

William B. Coley, MD (1909)[1]

OVERVIEW

1. The dual role of immune factors in fighting infection and stimulating cancer regression led to almost 70 years of treating inoperable cancers with "Coley's toxins", where severe infection cleared tumors in many cases.
2. A coordinated and balanced system of cytokine, interleukin and chemokine proteins is essential for normal cell development, metabolism and immune response.
3. Interleukin-2 (IL-2) and tumor necrosis factor-α (TNF) are induced during infection, but inflammatory and organ-damaging effects may continue even after the pathogen has been eliminated.
4. "Cytokine storm", a difficult-to-control and often lethal inflammation due to residual chemokines and cytokines and their induced proteins, is seen in sepsis, dengue shock syndrome (DSS) and after many infections, including Covid-19.
5. IL-2 supports the activation and survival of different types of T-cells that can induce antibody production by B-cells, tumor necrosis (lysis), or regulate tolerance and modulate the immune system.
6. However, treatment with recombinant forms of IL-2 can cause potentially lethal lung edema and other adverse events, thus inhibiting its anticancer use. Paradoxically, inhibiting IL-2 can also have lethal effects.
7. TNF regulates or is co-produced with an army of interleukins and chemokines (IL-1 family, IL-4, IL-6, IL-8, IL-17, IL-23 and others) that induce inflammation and fever in response to infection.
8. TNF, as its name implies, induces lysis of tumor cells. But its stimulation of autophagy, inflammatory proteins and proteases leads to unacceptable side effects, preventing direct use as a therapy.

DOI: 10.1201/9781003333319-7

9. TNF is overproduced in psoriasis, other autoimmune diseases or simply with age.
10. Inhibitors of TNF and related cytokines IL-17 and IL-23 are used clinically to control chronic diseases. However, those treated are immunosuppressed and more susceptible to infections.
11. Versions of IL-2 or TNF designed to be less toxic to normal cells are still being tested for cancer care despite many disappointments in clinical trials.

Authors note: This chapter is intended to give some explanation of the considerations for using treatments with cytokines or their inhibitors. It is not intended to replace a doctor's orders or to suggest self-prescribing. The marketing names for inhibitors of the cytokines/chemokines may have changed, especially for generic equivalents. Many will be replaced by molecules with similar or superior properties. Alternatively, they may have been withdrawn for unforeseen adverse events after long-term "phase 4" (post approval) clinical use.

COLEY'S TOXIN SHOWED THE INTERSECTION OF FACTORS FIGHTING INFECTION AND CANCER

In the waning decade of the 19th century, Dr. William B. Coley made a key observation: patients with untreatable cancers recovered after experiencing severe streptococcal skin infection (called erysipelas). Coley continued to explore the effect, by applying dead and growing streptococci, and streptococci mixed with other bacteria, especially *Serratia*, to patients with cancer, with variable success.[1] He observed, as others had before him, that patients who developed the worst bacterial infections in response to his applications often had the best antitumor response.

While Coley's ministrations might seem very dangerous to a modern observer, one should remember that cancer remained a mysterious, incurable disease until well into the 20th century. Doctors and patients were desperate for any therapy, even one that could result in a possibly fatal infection or at best an unsightly and painful rash. Coley published widely on using "Coley's toxins" for tumor regression, rejoicing in patients who survived the therapy and lived for decades (in one case, >43 years). He administered these until his death in 1936. Although studies have indicated that as many as 500 of the ~1000 patients Coley treated showed tumor regression, the preparations were difficult to standardize and apply. By the early 1960s, Coley's techniques gave way to chemotherapy, radiation, immunotherapy and targeted antitumor drugs.[2]

Yet Coley's work was not completely forgotten. His daughter, Helen Coley Nauts, helped found the Cancer Research Institute (see https://www.cancerresearch.org/en-us/blog/april-2015/what-ever-happened-to-coleys-toxins, accessed Feb. 7, 2023). Many researchers continued to refine Coley's methods and purify the active proteins we now call cytokines and chemokines from cellular supernatants. Breakthroughs came from using progressively purified bacterial components to treat cultivated human cells with isolated lipopolysaccharide (LPS) isolated from the bacterial cell walls instead of infectious bacteria, filtering the supernatant to remove extraneous matter and then concentrating the active fractions. With time, it was clear that there were indeed proteins produced by stimulated mammalian cells that could induce cancerous cells to lyse, at least in tissue culture. The most potent, and first to be cloned, are highlighted in this chapter.

ENTER THE CLONED AND PURIFIED MASTER REGULATORS

The conditionally toxic cytokines interleukin-2 (IL-2) and tumor necrosis factor (TNF) are important regulatory proteins present at low concentrations in normal sera until induced by infections, or, in the case of TNF, with aging. Similar to IFNs, both are involved in immune response to pathogens, have very high specific activity, on the order of 10^7–10^8 units/mg, and can contribute to an overactive immune response that complicates their use in therapies. Both IL-2 and TNF work at the top of cascades, controlling the production or activity of proteins of pathways with specific functions. Many proteins lower in the cascade are always present but only become activated when they are phosphorylated or otherwise changed by the enzymes above them. Figure 6.1 illustrates that these two different

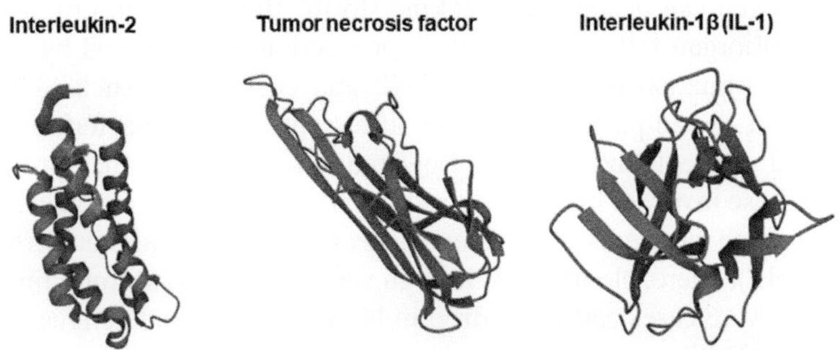

FIGURE 6.1 Structures of human IL-2 (from 1MA7.pdb), TNF (A chain of 1TN.pdb) and IL-1 (9ILB.pdb) illustrate different immune factor folding patterns. IL-2 is an α-helical bundle, a "cytokine-fold" also seen for the IFN family. TNF, in contrast, is a β-barrel, similar to other inflammatory cytokines such as IL-1.

cytokines have completely different structures. IL-2 is an α-helical bundle, similar to IFNs (see the previous chapter). TNF is almost exclusively β-strand, a characteristic it shares with the IL-1 family and related inflammatory proteins (such as IL-1β).

As discussed in the previous chapter, IFNs are vital for protection against pathogens. However, they can also stimulate autoimmune diseases such as systemic lupus erythematosus (SLE), either due to natural overproduction or long-term usage for treating diseases such as multiple sclerosis (MS). Similarly, both IL-2 and TNF have been implicated in diseases. Their inhibitors thus have established clinical uses.

A WEB OF CYTOKINES HAVE INFLAMMATORY AND ANTI-INFLAMMATORY ACTIVITIES

IL-2 and TNF are central molecules in a web of cell-produced cytokines and chemokines,[3] whose names, like those given to allergens (see Chapter 4), were assigned in order of their identification or cloning. Thus, interleukin number gives no obvious indication of interrelationship in structure and function.[4] For example, structural relatives of IL-1, discussed later as a fever-inducing cytokine, include IL-18 and IL-36. IL-18 in turn shares with IL-12 the ability to induce IFN-γ and natural killer (NK) cells. Further, many of the cytokines occur in different versions (isoforms) that may have their own unique activities. A complex web of cytokines, chemokines and enzymes can have both autocrine (acting on the cells that produce them) and paracrine (acting on other cells) effects (Figure 6.2).

Storming cytokines. While inflammation and fever are part of a normal immune response to a bacteria or virus, they may also contribute to tissue damage. What may start out as an appropriate immune response to a pathogen can become a "cytokine storm" that can even progress to lethal multiorgan failure.[5] After their production is stimulated by infections, inflammation from residual TNF and chemokines such as IL-4, IL-6 and IL-8 can persist, especially after sepsis (where bacteria invade the blood and colonize tissues). Lowered expression of other proteins, such as those involved in regulating vascular permeability,[6] can also cause post-infection problems, as is seen for example in dengue hemorrhagic fever (DHF).[7] Here, blood leaks through vessel walls into surrounding tissues, in severe cases causing a drop in blood pressure and dengue shock syndrome (DSS).

Some have suggested that since little measurable virus is found, and human proteins and immune factors participate in changing the permeability of the vessel walls that leads to hemorrhage, DHF and tissue damage from other viruses should be called "virus induced cytokine (or

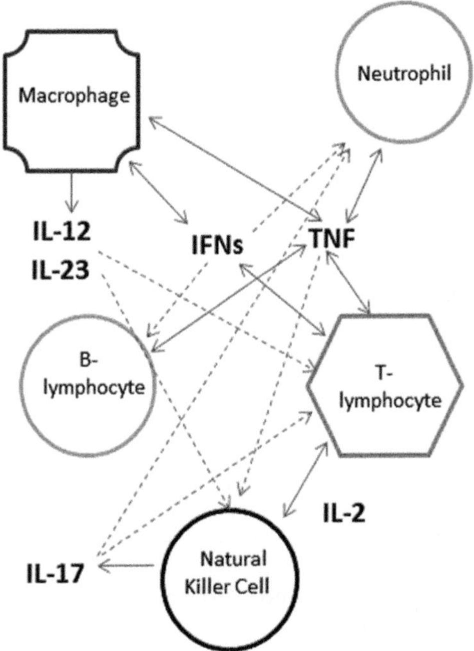

FIGURE 6.2 Cytokine web. IFNs, TNF-α and various interleukins discussed in Chapters 5 and 6, are produced by different types of cells (solid arrows). These affect both the cells that produce them (double-headed arrows) and other cells (dashed arrows). For a more complete overview of the many interactions of cytokines and chemokines with the cells that produce them, see https://www.bosterbio.com/media/pdf/collaterals/pathways/Cytokine_Network_Pathway.jpg (accessed Aug. 9, 2023).

chemokine) storm" to differentiate it from that seen in sepsis caused by other infections where, for example, bacteria or their toxins continue to cause tissue damage. DHF typically develops in those who have previously had an infection with another form (serotype) of the dengue virus.[8] Their antibodies against the first form of dengue are too weak to inhibit the second serotype of the virus; they may even help the second virus enter cells and grow to titers 10–100× higher, as is typical of patients who develop DHF. Hundreds of children, who are particularly susceptible, died in recent outbreaks in Brazil. The World Health Organization (WHO) estimates 4000 deaths occur per year due to DHF, mostly in Asia.

The same storming cytokines have been implicated in bacterial and other severe viral illnesses, most recently Covid-19.[9] However, paradoxically, immune suppression, including the lack of proper induction of IFNs after infection, may also lead to severe illness.[10]

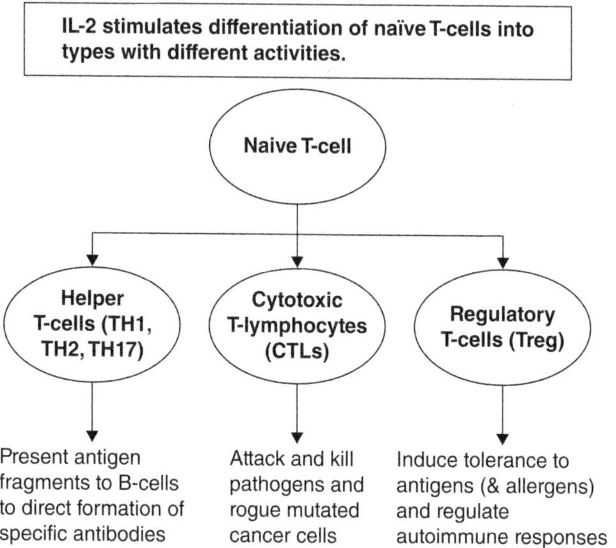

FIGURE 6.3 IL-2 stimulates T-cell differentiation into different classes of mature cells with different activities.

The ups and downs of Interleukin 2 (IL-2) is an ongoing immunological saga. This "Type 1" cytokine, typically co-produced with IFN-γ, was first identified as a protein needed to keep T-cells (lymphocytes) growing in cell culture. This was important for the original method for producing IFN from human blood cells in mixed lymphocyte culture (see Chapter 5). One of IL-2's early names was "T-cell replacing factor", meaning it extended the growth of lymphocytes in the cultured buffy coats stimulated with phorbol esters. IL-2 activates T-lymphocytes to develop and differentiate into cells with many different properties, including cytotoxic T-lymphocytes (CTLs), which actively kill pathogens and tumor cells, or helper T-cells, which stimulate B-cells to produce antibodies. IL-2 controls this diversity of effects by interacting with its high- or low-affinity receptors on the surface of cells and cellular phosphatases and kinases that control the concentration of phosphoinositides and other small molecules (Figure 6.3 shows that IL-2 induces differentiation of T-cells to perform different tasks).[11,12]

IL-2 IS NECESSARY FOR DEVELOPMENT AND DEVELOPING SELF-TOLERANCE

Eliminating the gene for IL-2 in mice illustrated the very important role of this cytokine in development. IL-$2^{-/-}$ mice were born and developed normally for three to four weeks (meaning that unlike insulin, absence of

IL-2 was not embryonically lethal). However, about half died before they were nine weeks old, and the rest developed inflammatory bowel disease resembling ulcerative colitis in humans. Two surprises showed how different IL-2's activities in a whole organism were from those seen with cultured cells. First, although IL-2 stimulates B-cells to make antibodies in cultured mixed lymphocytes, mice unable to make functional IL-2 had normal thymus development but overproduced IgE and IgG antibodies! This revealed IL-2's central role in controlling autoimmunity by stimulating regulatory T-cells (or Tregs).[13] On the positive side, mice who could not make IL-2 did not show the "cytokine storm" that follows injection with LPS in normal mice. Thus, IL-2 could rein in the immune system, while overreaction could be lethal, as attempts to use recombinant IL-2 in therapy would show (see the next section).

IL-2 causes lung edema. Based on its ability to induce T-lymphocytes (CTLs) that attack tumor cells, pure, recombinant IL-2 was first tested in the 1980s for cancer therapy. Injections in some patients caused remission, and even total clearing of tumor cells. The injections indeed enhanced CTLs, and cytokines such as TNF and IFN-ɣ (which slows cell growth), leading in some cases to tumor remission. Although a few patients responded extraordinarily well, with metastatic cells throughout the body cleared, IL-2 rapidly revealed itself to be an extremely toxic protein. These early experiments demonstrated the low therapeutic index (ratio of active to toxic concentration) of IL-2; the dose required to kill cancer cells proved perilously close to a lethal one for the patient. Treatments were accompanied by fever and/or cardiac (heart), renal (kidney), neurologic, GI or hepatic (liver) dysfunction, and a potentially fatal capillary leak syndrome such as that described above for DSS. Lung edema and capillary leak syndrome were also seen when even tiny amounts of IL-2 were injected into mice. The immune system, perhaps due to overproduction of TNF and IFN-ɣ, appeared to be raging out of control.

IL-2 in cancer immunotherapy. Despite extensive results showing toxic effects in murine studies, some studies continued in patients with advanced cancers. Multistep approaches, combining IL-2 with newer immune-directed therapies, were tested in those whose cancers were not vanquished by repeated chemotherapy. It was combined in a triple treatment that led to eradication of tumors and survival >22 months of a patient with metastatic breast cancer, in a combination of genetic and immunotherapy.[14] First, her genome was sequenced to identify possible tumor-driving mutations in four proteins (SLC3A2, KIAA0368, CADPS2 and CTSB) in the rapidly growing cancer cells. Then the patient's own cells were reprogrammed *in vitro* to generate tumor-infiltrating lymphocytes (TILs) specific for the four mutant proteins. The TILs were stimulated to grow in culture and reinfused into the patient along with IL-2 (aldesleukin, 720,000 U/kg, every eight hours for seven doses "to tolerance"). The third part of the therapy was "checkpoint blockade" using the monoclonal antibody pembrolizumab (Keytruda, 2 mg/kg, at day – 2, 21, 42, 63), which inhibits a protein, PD-L1, that normally inhibits T-cell responses. Pembrolizumab lifts this blockade, allowing NK and CTLs to attack the tumor.

In that important last step of the above example, pembrolizumab worked with IL-2 to remove two brakes on the T-cells, a two-step activation of T-cells used because only about 15% of patients in other studies achieved complete response to pembrolizumab alone. Adding IL-2 for double activation may be needed because NK and other T-cells can take up proteins, including the PD-L1 receptor, from tumor cells they encounter, a process called trogocytosis.[15] Once they themselves express the receptor, the NK cells no longer attack the PD-L1-positive tumor cells.

More complete results of the triple treatment study will be available in four to five years. However, comments from one of the study directors indicate that even this severe combined therapy has a low response rate and the reported recovery may be a "medical unicorn". On the other hand, genetic studies can now indicate patients who may be more suited to immunotherapies such as that described.[16]

REDIRECTING AND INHIBITING IL-2

Of course, the major way to prevent adverse events in such immune-stimulated therapies would be to keep IL-2 from affecting normal cell pathways and concentrate its effects solely on tumors. One way is to use IL-2 to induce CTLs from blood cells removed from the body. However, removing and replacing a patient's cells leads to other problems, including swelling at the injection site and infection. Another approach is to apply very low-dose IL-2 subcutaneously, as tested for alopecia aureate, or autoimmune induced pattern balding.[17] IL-2-induced Tregs could tame the immune response leading to balding. However, patients had side effects. Other studies indicate that IL-2 and other cytokines may be overproduced in alopecia,[18] suggesting that the hair loss is due to pathway dysregulation that will not easily be solved by adding a single protein.

A more recent approach is to change the structure of IL-2 to restrict its binding only to low-affinity receptors, inducing only tumor cell lysis. "THOR-207" contains an unnatural amino acid (i.e., not one of the 20 shown in Table 3.1) that prevents binding to the high-affinity α chain, CD25, of the IL-2 receptor on T-cells. The modified protein can still stimulate NK cells through binding to the lower affinity β and γ receptor subunits. Another approach is to link IL-2 directly to a protein that blocks its activities. The three-part molecule (IL-2/linker/inhibitor) cannot bind to receptors on normal cells, preventing it from activating NK cells in the periphery and inducing toxic effects. Once the drug gets to the microenvironment of tumor cells, the overproduced proteases cleave the linker, releasing IL-2 from the inhibitory domain.[19] Modified IL-2s have (as of early 2023) not achieved the desired endpoints of their clinical trials, but other attempts continue.

Inhibitors of IL-2 also have mixed success in the clinic

While direct treatment with IL-2 continues to struggle with side effects, inhibitors of IL-2 have been used for some time. The earliest of these is basiliximab (trade name Simulect), which was approved by the FDA in 1998 to prevent graph versus host reactions after kidney and other transplants. Basiliximab is a chimeric mouse-human monoclonal antibody that binds the high-affinity α chain, CD25, of the IL-2 receptor on T-cells, preventing IL-2 from binding and inducing inflammatory cytokines. The antibody prevents stimulation of T-cells to replicate, lowering their ability to induce B-cells to produce antibodies against the foreign graft.

A later antibody to the IL-2 receptor, Daclizumab, was approved in 2016 for the treatment of progressive MS in patients who had not responded to at least one first-line treatment (often a high-dose steroid infusion). It was withdrawn due to severe adverse events in 2018.[20] It is however important to note that while this antibody prevented IL-2 from binding to its high-affinity receptor, it increased the soluble IL-2 levels, which could still stimulate NK cells through binding to the β and γ receptor subunits.

The next storming cytokine, TNF, can also have opposing activities.

TNF, ONE OF THE MAJOR FACTORS IMPLICATED IN CYTOKINE STORM

This is an excellent example of a conditional toxin. Its history illustrates how destructive a single protein can be when overproduced or taken out of context in the body. Many researchers have been frustrated by this potent molecule, which, as its name implies, can cause tumors to stop growing and undergo a pathway to necrosis. Indeed, TNF was first isolated as a cytotoxin (i.e., a protein toxic to cells) produced by human lymphocytes stimulated by lipopolysaccharides (LPS). LPS, also called endotoxin, are fragments of bacterial cell wall released into the blood stream when bacteria lyse. LPS is immediately detected by the immune system as a marker of infection. TNF's most important natural role is to control invading pathogens' growth. Its ability to also cause apoptosis and necrosis of tumor cells may be a secondary activity. Many of TNF's pleiomorphic (many sided) effects are due to differential binding to a whole family of receptors (referred to as

TNFR structure family (TNFRSF) members). As with IL-2, pharmaceutical companies are seeking mutated forms of TNF to differentially affect these receptors.[21]

The positive and negative effects of TNF became clearer after cloning

The genes for TNF and related proteins were cloned in the 1980s, allowing them to be purified in large amounts and permitting the first clinical trials. Cloning revealed that many TNF family members had other names consistent with activities in many different assays and syndromes. One name was cachectin, the factor that induced wasting and appetite loss (cachexia) in individuals with infectious diseases or cancer. TNF shared sequence identity and receptors with lymphotoxins (for example, TNF-β, now called lymphotoxin, LTα), which also induced cancer cell death. Early studies indicated purified TNF-induced efficient lysis of cultured cancerous, but not normal human cells.

TNF was (for a short time) proclaimed the new magic bullet against cancer. Unfortunately, although TNF did indeed induce necrosis in cancer cells, it was obvious that injection of TNF in people could cause inflammation, wasting and, if used in high enough amounts to kill tumors, fatal septic shock. The therapeutic index of the protein was small, and side effects included cytokine storm. There have been sporadic successes with injecting TNF directly into a tumor, but the safety profile prevented its widespread use.[22,23]

Further, TNF can even cause cancer or contribute to its spread. TNF signaling exacerbated human head and neck squamous cell carcinoma.[24] Mice without the TNF gene had few apparent abnormalities and were immune to some artificially induced tumors! The $TNF^{-/-}$ mice lost weight at low doses of TNF; higher doses could be lethal.

The precarious balance of TNF activation cascades can account for diverse activities

As Figure 6.4 shows, the activities of TNF require balancing among several possibilities. TNF can determine cell fate by binding to receptor 1, which contains a "death domain". This stimulates a series of molecules ("the TRADD-TRAF2-RIP-TAK1-IKK pathway") that leads to activation (in this case, phosphorylation) of NFκB. Most simply, one can picture TNF sitting atop a series of balls (representing kinases) poised on a hill, able to roll in several directions. Each kinase can activate (phosphorylate) a protein lying below it, in one direction leading to the phosphorylation and

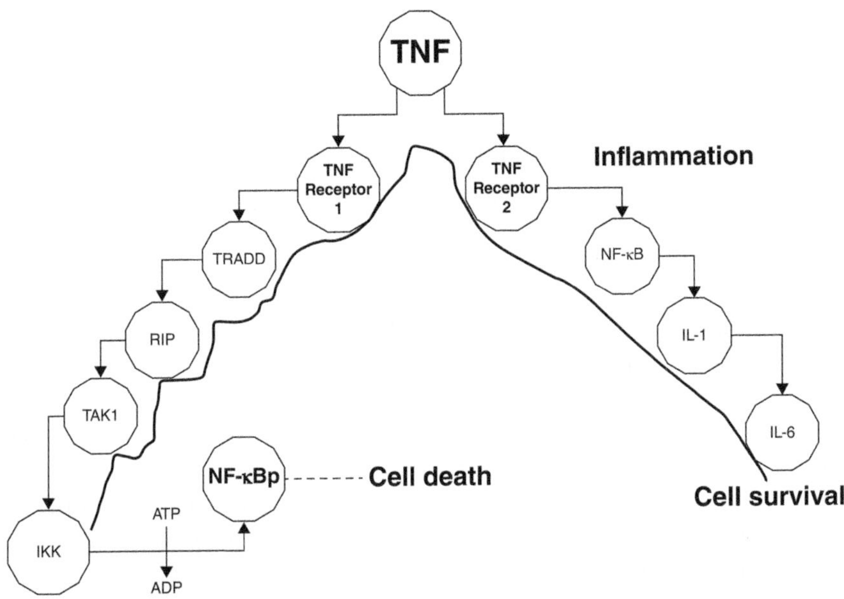

FIGURE 6.4 TNF is at the top of signaling pathways that can lead to cell death or survival with induction of inflammatory proteins. Binding to receptor 1, which contains a death domain (DD), can start a cascade of kinases (TRADD-TRAF2-RIP-TAK1-IKK pathway) leading to NF-κB phosphorylation to initiate apoptosis/necrosis in cancer cells. Binding to receptor 2 (which has no DD) leads to cell survival and induction of inflammatory cytokines that may play a role in infection control but can cause human diseases and heart failure[33] when TNF is chronically expressed. Abbreviations: TNF receptor-associated death domain (TRADD), TNF receptor-associated factor (TRAF)2, receptor-interacting protein (RIP), transforming growth factor-beta-activated kinase 1 (TAK1), IKK: I kappa kinase, IL-1, IL-6.

activation of NFκB (Figure 6.4). Once this nuclear transcription factor is activated, it can turn on genes to produce proteases and nucleases that mediate the self-destruction (autophagy) of the cell (by apoptosis or necrosis, depending on the factors involved).[25] If TNF rolls in the other direction, it stimulates other proteins more conducive to cell survival.

A missing factor in the TNF cascade can lead to a lack of response. In a normal cell, NfκB is turned off by its specific inhibitor when the infection is dealt with, and the genes it controls are downregulated. But in the continuous presence of TNF, or one of its coworkers (IL-17, IFN-γ, IL-23[26]), the pathway remains turned on.

TNF is overproduced in several diseases. After the TNF gene was cloned, and one could accurately measure the levels in the blood, high levels were found during bacterial infections, particularly seen with those that cause tuberculosis. TNF is particularly high when bacteria invade the blood, in a condition called sepsis, where it is associated with blood vessel seepage and edema around the heart. TNF's inflammatory effects are exacerbated by the toxins produced by bacteria that cause sepsis, as will be discussed in a later chapter. Confusingly, treatments with inhibitors of TNF have not shown great effects in mitigating septic shock and reducing patient mortality. TNF may not be the real "bad guy" in sepsis, as there are plenty of other persistent inflammatory cytokines that could be blamed (i.e., targeted for inhibitor design).

"Pyrokines" ("fever cytokine"), IL-1 (α [IL-1F1] or β-[IL-1F2]), control many of the perceived negative aspects of the inflammatory response. Pyrogenic (fever-inducing) effects could be part of the protective response against infection, however miserable they may make the patient feel. There are 11 different members of the IL-1 family, many of whose genes occur near one another on chromosome 2. These fall into three to four distinct sequence families with a common fold[4,27] (see Figure 6.1). The family includes auto-inhibitors, including IL-1Ra (IL-1F3), which blocks binding of IL-1α to its receptor. The presence of these natural inhibitors indicates how destructive IL-1 family members can be, if expressed on their own. Mice overexpressing TNF spontaneously develop RA, which can be prevented by injecting IL-1Ra. Similarly, IL-1Ra injections have had some effect in reducing inflammation in human RA patients. IL-1 overexpression has been linked in at least one patient with the epidermal disorder, Sweet syndrome. But while TNF may play a protective role against infection, it is conditionally destructive elsewhere.

TNF is overexpressed in autoimmune disorders and aging, with increased levels in those over 60. Higher levels of TNF can even predict shorter life expectancy. Infections in the elderly may not cause further rise in TNF levels, suggesting that they are resistant to its effects. High levels of TNF are found in patients of all ages with chronic diseases, including diabetes, cancer, osteoporosis, MS, and inflammatory bowel diseases. TNF overproduction marks two of the most common autoimmune disorders: rheumatoid arthritis (RA) and psoriasis (especially psoriatic arthritis). It is estimated that RA affects about 1% of the world population, causing irreversible damage to joints and many co-morbidities. Severe psoriasis affects about 7.5 million Americans. Elevated TNF and related cytokines are also found in hidradenitis suppurativa (HS) and inflammatory bowel diseases.[28]

Interest thus shifted from using TNF directly to treat cancer to developing inhibitors. These have proved to be very useful for patients resistant to long-term treatments with steroids, coal-tar-based creams and NSAIDs.

Inhibitors of TNF treat disease. Many inhibitors of TNF are now used to treat RA, psoriasis, HS and Crohn's disease. Inhibitors of TNF (including the monoclonal anti-TNF antibodies infliximab (Remicade®) adalimumab (Humira®), golimumab (Simponi Aria) and certolizumab pegol along with etanercept, (a TNF-receptor/IgG-Fc fusion protein) can be miracle drugs for those suffering from systemic and psoriatic arthritis* or RA. Golimumab, which binds directly to TNF and neutralizes it, gave an 82% response rate in patients with early peripheral spondyloarthritis (SpA), with 53% still in remission 18 months after the drug was stopped.[29]

Going further down the pathway to other cytokines co-produced with TNF can give even better responses. Antibodies targeting interleukin (IL)-12/23 and IL-17 may give much faster skin clearance of severe psoriasis. These include ustekinumab, an IL-12/23 inhibitor; abatacept, a CTLA4-Ig inhibitor and smaller drugs, tofacitinib, a JAK inhibitor; and apremilast, a phosphodiesterase-4 inhibitor.[30] Other biologic agents for psoriatic arthritis are in clinical trials.

Besides the proteins highlighted in this chapter, IL-2 and TNF, other cytokines have also been linked to the cytokine storm phenomenon. Dupilumab (Dupixent), a fully human monoclonal antibody, blocks the shared receptor component for IL-4 and IL-13, cytokines that are key and central drivers of multiple Type 2 diseases. Dupilumab is approved for patients with Type 2 inflammatory diseases, including atopic dermatitis, asthma[31] and chronic rhinosinusitis with nasal polyps.

Paradoxically, treating with TNF inhibitors may also lead to psoriasis! In this case, switching to another drug that inhibits two other cytokines, IL-12 and IL-23 (with topical corticosteroids and ustekinumab antibody), may help. The success of using these protein inhibitors (adalimumab, etanercept, infliximab, secukinumab and ustekinumab) is evidenced by the recent introduction of "biosimilars" to the market, including Benepali for Enbrel and Remsima for Remicade. Although these biosimilars are currently not substantially less expensive, competition among drugmakers should eventually reduce the cost for using these treatments. This may lead to their use (over much cheaper steroids) earlier in psoriasis treatment and to treat milder cases. The multiple different effective treatments for this debilitating condition will pay for themselves in terms of enhanced ability of recovered patients to live and work without pain.

---

* A patient, who had been treated since early childhood with steroids/coal tar creams and shampoos, called her anti-TNF antibody treatment a miracle, as the skin lesions that covered most of her body disappeared. Joint damage due to psoriatic arthritis was unfortunately not reversed, emphasizing the need for treatment with direct inhibitors as early as possible.

TAMING THE IMMUNE SYSTEM WITH THESE INHIBITORS DECREASES RESISTANCE TO INFECTION

Unfortunately, many of the treatments that block TNF and related cytokines also reduce resistance and immunity, as one also eliminates the positive effects of TNF and its related cytokines. Those taking these inhibitors are more susceptible to serious infections, such as *Streptococcal* toxic shock syndrome or invasive fungal disease. While many individual proteins are overproduced in autoimmune diseases, their inhibitors can also diminish the ability of these proteins to perform their normal functions. The antibodies must be administered as a shot or an infusion, providing another barrier to their routine use.

The major risk associated with TNF and IL-2 inhibitors is increased danger of severe infections or patients developing antibodies or allergies to the infusion (such as those discussed in Chapters 2 and 4). Thus, their use must be reserved for those with defined overproduction and serious disease. All those taking these drugs must also be aware of their impaired immune function and encouraged to self-protect against infections, by avoidance of crowds and appropriate masking.

RESTORING BALANCE WITH INHIBITORS OF INFLAMMATORY CYTOKINES IS COMPLICATED

So how does one decide among new therapies and what degree of side effects are acceptable? Each patient's psoriasis may have different causes. A cytokine panel may suggest which of the known agents are overproduced. However, many of these factors are produced locally in the body, and the production of one may be dependent on another, so a single measurement may not implicate a particular cytokine. Thus, a doctor may choose to treat with one inhibitor, and then switch a patient to another if they are not satisfied with their progress. They may also choose to increase or reduce the dose for drugs. Even inhibitors that target the same cytokine can have different effects. For two inhibitors of IL-23, ustekinumab had the highest drug survival (time to treatment discontinuation). Despite a 100% response rate (improvement in psoriasis), secukinumab had the lowest drug survival of those compared and the most related adverse events (mostly infections). These results illustrate the dual nature of factors in the TNF pathway: they are positive in dealing with infections, negative when chronically overproduced in RA and psoriasis.

Are inflammatory cytokines essential? The answer has to be: it depends. In Chapter 3, the amino acids that make up proteins were classed as essential, non-essential or conditionally essential based on whether the body could meet its requirements without direct uptake from foods.

Obviously, all amino acids are essential in that they are required for producing proteins the body needs to survive. Just as there have been difficulties determining whether glutamine is even conditionally essential, there is also a problem determining which proteins made by the human body can be safely eliminated if they are shown to be elevated in disease states.

The most obvious way to determine whether a protein is essential is to eliminate the gene for it and determine whether the organism develops and is healthy. Deletion of the gene for insulin is lethal at the earliest stages of fetal development. Similarly, most of the genes for producing blood cells, tissues and bone, or adding glycosylation to proteins, cannot be eliminated or result in severely damaged offspring. Less lethal mutations result in muscular dystrophy, cystic fibrosis and a multitude of rarer diseases.

The dual nature of cytokines is also illustrated by the IL-1 family members, where removing the naturally occurring inhibitor, IL-1Ra, has more effect on development than eliminating any of the other members, including the fever inducer IL-1α. Deleting the IL-1Ra gene yields stunted mice with inflamed, damaged arteries, which spontaneously develop RA and do not reproduce. Mice lacking IL-1α, β or IL-18 develop normally, do not develop cancers and live a normal lifespan but are more susceptible to infections.

In the unsanitary and antibiotic-free world of our ancestors, in which only the fittest babies survived, factors that stimulated the immune response against pathogenic organisms were essential. Children completely lacking IL-2 or TNF succumb to infection in early life; survivors develop irritable bowel syndrome. Severely ill, ICU patients made less TNF and two other inflammatory cytokines, IL-1 and IL-6,[32] than normal controls when their leucocytes were stimulated with LPS.

Thus, the very proteins that make one feel the worst, that are overproduced with chronic diseases and aging, are needed for protection against infections. ERAP and other enzymes that today are known to be overproduced in autoimmune diseases may also have protected against the bubonic plague, as described in Chapter 1. This is the tradeoff with conditionally toxic proteins.

CONCLUSIONS

To sum up, IL-2 and TNF both play important roles in the immune response to infections but have side effects that have limited their use in cancer therapy. They are both probably essential in their own way: mice unable to make IL-2 develop autoimmune diseases and die of ulcerative bowel syndrome, probably due to removal of the modulatory control of IL-2 on the immune system via its induction of Tregs (which also play a role in inducing tolerance to allergens). Inhibiting IL-2 through an antibody to one part of its receptor can have lethal consequences, as can subcutaneous injection of IL-2. Low-dose IL-2 may enhance Tregs in alopecia aureate, but its other effects may even induce baldness.

TNF is an essential protein for protecting against pathogens (most notably tuberculosis and other mycobacteria). It also helps in eliminating stray cells before they can cause tumor formation, as illustrated by the enhanced development of certain cancers in those taking TNF inhibitors. Some proteins in the TNF family can protect the heart against ischemic injury. But TNF production is higher in older individuals and is increased even in young patients with RA and psoriasis. While toxic effects have limited the uses of either cytokine directly, inhibitors of TNF and other related inflammatory cytokines, overproduced in autoimmune disease, are now used routinely in the clinic to treat RA, psoriasis and Crohn's disease.

While IL-2, TNF and other cytokines mediate the immune system's response to invaders, directing T-cell activation (IL-2) and pathogen lysis (TNF, its series of receptors and related cytokines), they can also cause a continuing storm of symptoms even after infection has been defeated. Rampaging cytokines can lead to severe disease and even death. They are thus normally tightly controlled in the body.

The anticancer activities of TNF and IL-2 may have developed secondarily to their primary role in protecting against infections, during the ongoing microbial wars further discussed in Chapter 8. Newer forms of either cytokine may be able to dissociate their many activities, to eliminate side effects and partner them with other immunotherapies. Similarly, ideal regulators of TNF-related cytokines would control inflammation, while still protecting against infections. Better therapeutics in the future may bind specifically to only one cellular receptor or target other proteins in the cytokine web.

REFERENCES

1. Coley WB. The treatment of inoperable sarcoma by bacterial toxins (the mixed toxins of the *Streptococcus erysipelas* and the *Bacillus prodigiosus*). Proc R Soc Med. 1910;3(Surg Sect):1–48. PubMed PMID: 19974799; PMCID: PMC1961042.
2. Schein CH. Repurposing approved drugs for cancer therapy. Br Med Bull. 2021;137(1):13–27. doi: 10.1093/bmb/ldaa045. PubMed PMID: 33517358; PMCID: PMC7929227.
3. Lissoni P, Messina G, Pelizzoni F, Rovelli F, Brivio F, Monzon A, Crivelli N, Lissoni A, Tassoni S, Sassola A, Pensato S, Di Fede G. The fascination of cytokine immunological science. J Infect Epidemiol. 2020;3(1):18–28. doi: 10.29245/2689-9981/2020/1.1155. PubMed PMID: 29705848.
4. Schein CH. The shape of the messenger: using protein structure information to design novel cytokine-based therapeutics. Curr Pharm Des. 2002;8(24):2113–29. doi: 10.2174/1381612023393161. PubMed PMID: 12369857.

5. Fajgenbaum DC, June CH. Cytokine storm. N Engl J Med. 2020;383(23): 2255–73. doi: 10.1056/NEJMra2026131. PubMed PMID: 33264547; PMCID: PMC7727315.
6. Garishah FM, Boahen CK, Vadaq N, Pramudo SG, Tunjungputri RN, Riswari SF, van Rij RP, Alisjahbana B, Gasem MH, van der Ven A, de Mast Q. Longitudinal proteomic profiling of the inflammatory response in dengue patients. PLoS Negl Trop Dis. 2023;17(1):e0011041. Epub 20230103. doi: 10.1371/journal.pntd.0011041. PubMed PMID: 36595532.
7. Green S, Vaughn DW, Kalayanarooj S, Nimmannitya S, Suntayakorn S, Nisalak A, Lew R, Innis BL, Kurane I, Rothman AL, Ennis FA. Early immune activation in acute dengue illness is related to development of plasma leakage and disease severity. J Infect Dis. 1999;179(4):755–62. doi: 10.1086/314680. PubMed PMID: 10068569.
8. Bowen DM, Lewis JA, Lu W, Schein CH. Simplifying complex sequence information: a PCP-consensus protein binds antibodies against all four Dengue serotypes. Vaccine. 2012;30(42):6081–7. Epub 20120731. doi: 10.1016/j.vaccine.2012.07.042. PubMed PMID: 22863657; PMCID: PMC3439202.
9. Frisoni P, Neri M, D'Errico S, Alfieri L, Bonuccelli D, Cingolani M, Di Paolo M, Gaudio RM, Lestani M, Marti M, Martelloni M, Moreschi C, Santurro A, Scopetti M, Turriziani O, Zanon M, Scendoni R, Frati P, Fineschi V. Cytokine storm and histopathological findings in 60 cases of COVID-19-related death: from viral load research to immunohistochemical quantification of major players IL-1beta, IL-6, IL-15 and TNF-alpha. Forensic Sci Med Pathol. 2022;18(1):4–19. Epub 20210831. doi: 10.1007/s12024-021-00414-9. PubMed PMID: 34463916; PMCID: PMC8406387.
10. Remy KE, Mazer M, Striker DA, Ellebedy AH, Walton AH, Unsinger J, Blood TM, Mudd PA, Yi DJ, Mannion DA, Osborne DF, Martin RS, Anand NJ, Bosanquet JP, Blood J, Drewry AM, Caldwell CC, Turnbull IR, Brakenridge SC, Moldwawer LL, Hotchkiss RS. Severe immunosuppression and not a cytokine storm characterizes COVID-19 infections. JCI Insight. 2020;5(17). Epub 20200903. doi: 10.1172/jci.insight.140329. PubMed PMID: 32687484; PMCID: PMC7526441.
11. Ross SH, Cantrell DA. Signaling and function of interleukin-2 in T lymphocytes. Annu Rev Immunol. 2018;36:411–33. doi: 10.1146/annurev-immunol-042617-053352. PubMed PMID: 29677473; PMCID: PMC6472684.
12. Braun W, Schein CH. Membrane interaction and functional plasticity of inositol polyphosphate 5-phosphatases. Structure. 2014;22(5):664–6. doi: 10.1016/j.str.2014.04.008. PubMed PMID: 24807076.
13. Malek TR. The biology of interleukin-2. Annu Rev Immunol. 2008;26:453–79. doi: 10.1146/annurev.immunol.26.021607.090357. PubMed PMID: 18062768.
14. Zacharakis N, Chinnasamy H, Black M, Xu H, Lu YC, Zheng Z, Pasetto A, Langhan M, Shelton T, Prickett T, Gartner J, Jia L, Trebska-McGowan K, Somerville RP, Robbins PF, Rosenberg SA, Goff SL, Feldman SA. Immune recognition of somatic mutations leading to complete durable regression in metastatic breast cancer. Nat Med. 2018;24(6):724–30. Epub 20180604. doi: 10.1038/s41591-018-0040-8. PubMed PMID: 29867227; PMCID: PMC6348479.

15. Hasim MS, Marotel M, Hodgins JJ, Vulpis E, Makinson OJ, Asif S, Shih HY, Scheer AK, MacMillan O, Alonso FG, Burke KP, Cook DP, Li R, Petrucci MT, Santoni A, Fallon PG, Sharpe AH, Sciume G, Veillette A, Zingoni A, Gray DA, McCurdy A, Ardolino M. When killers become thieves: Trogocytosed PD-1 inhibits NK cells in cancer. Sci Adv. 2022;8(15):eabj3286. Epub 20220413. doi: 10.1126/sciadv.abj3286. PubMed PMID: 35417234; PMCID: PMC9007500.
16. Schein CH. Distinguishing curable from progressive dementias for defining cancer care options. Cancers. 2023;15(4):1055. PubMed PMID: doi:10.3390/cancers15041055.
17. Le Duff F, Bouaziz JD, Fontas E, Ticchioni M, Viguier M, Dereure O, Reygagne P, Montaudie H, Lacour JP, Monestier S, Richard MA, Passeron T. Low-dose IL-2 for treating moderate to severe alopecia areata: a 52-week multicenter prospective placebo-controlled study assessing its impact on T regulatory cell and NK cell populations. J Invest Dermatol. 2021;141(4):933–6 e6. Epub 20200914. doi: 10.1016/j.jid.2020.08.015. PubMed PMID: 32941917.
18. Waskiel-Burnat A, Osinska M, Salinska A, Blicharz L, Goldust M, Olszewska M, Rudnicka L. The role of serum Th1, Th2, and Th17 cytokines in patients with alopecia areata: clinical implications. Cells. 2021;10(12). Epub 20211202. doi: 10.3390/cells10123397. PubMed PMID: 34943905; PMCID: PMC8699846.
19. Ortolano N. Cancer vs. predator. Drug Discovery News 2022;18(2):26–7.
20. Rommer PS, Berger K, Ellenberger D, Fneish F, Simbrich A, Stahmann A, Zettl UK. Management of MS patients treated with daclizumab – a case series of 267 patients. Front Neurol. 2020;11:996. Epub 20200908. doi: 10.3389/fneur.2020.00996. PubMed PMID: 33013658; PMCID: PMC7506133.
21. Mahoney KM, Rennert PD, Freeman GJ. Combination cancer immunotherapy and new immunomodulatory targets. Nat Rev Drug Discov. 2015;14(8):561–84. doi: 10.1038/nrd4591. PubMed PMID: 26228759.
22. Libert C, Wielockx B, Hammond GL, Brouckaert P, Fiers W, Elliott RW. Identification of a locus on distal mouse chromosome 12 that controls resistance to tumor necrosis factor-induced lethal shock. Genomics. 1999;55(3):284–9. Epub 1999/03/02. doi: S0888-7543(98)95677-4 [pii] 10.1006/geno.1998.5677. PubMed PMID: 10049582.
23. Ruddle NH. Lymphotoxin and TNF: how it all began—a tribute to the travelers. Cytokine Growth Factor Rev. 2014;25(2):83–9. Epub 20140212. doi: 10.1016/j.cytogfr.2014.02.001. PubMed PMID: 24636534; PMCID: PMC4027955.
24. Jackson-Bernitsas DG, Ichikawa H, Takada Y, Myers JN, Lin XL, Darnay BG, Chaturvedi MM, Aggarwal BB. Evidence that TNF-TNFR1-TRADD-TRAF2-RIP-TAK1-IKK pathway mediates constitutive NF-kappaB activation and proliferation in human head and neck squamous cell carcinoma. Oncogene. 2007;26(10):1385–97. Epub 20060904. doi: 10.1038/sj.onc.1209945. PubMed PMID: 16953224.
25. Holbrook J, Lara-Reyna S, Jarosz-Griffiths H, McDermott M. Tumour necrosis factor signalling in health and disease. F1000Res. 2019;8. Epub

20190128. doi: 10.12688/f1000research.17023.1. PubMed PMID: 30755793; PMCID: PMC6352924.

26. Ehst B, Wang Z, Leitenberger J, McClanahan D, De La Torre R, Sawka E, Ortega-Loayza AG, Strunck J, Greiling T, Simpson E, Liu Y. Synergistic induction of IL-23 by TNFalpha, IL-17A, and EGF in keratinocytes. Cytokine. 2021;138:155357. Epub 20201102. doi: 10.1016/j.cyto.2020.155357. PubMed PMID: 33153894; PMCID: PMC7856048.
27. Fields JK, Gunther S, Sundberg EJ. Structural basis of IL-1 family cytokine signaling. Front Immunol. 2019;10:1412. Epub 20190620. doi: 10.3389/fimmu.2019.01412. PubMed PMID: 31281320; PMCID: PMC6596353.
28. Savage KT, Flood KS, Porter ML, Kimball AB. TNF-alpha inhibitors in the treatment of hidradenitis suppurativa. Ther Adv Chronic Dis. 2019;10: 2040622319851640. Epub 20190527. doi: 10.1177/2040622319851640. PubMed PMID: 31191873; PMCID: PMC6540495.
29. Carron P, Varkas G, Renson T, Colman R, Elewaut D, Van den Bosch F. High rate of drug-free remission after induction therapy with golimumab in early peripheral spondyloarthritis. Arthritis Rheumatol. 2018;70(11):1769–77. doi: 10.1002/art.40573. PubMed PMID: 29806090.
30. Schein CH. Repurposing approved drugs on the pathway to novel therapies. Med Res Rev. 2020;40(2):586–605. Epub 20190820. doi: 10.1002/med.21627. PubMed PMID: 31432544; PMCID: PMC7018532.
31. Sher LD, Wechsler ME, Rabe KF, Maspero JF, Daizadeh N, Mao X, Ortiz B, Mannent LP, Laws E, Ruddy M, Pandit-Abid N, Jacob-Nara JA, Gall R, Rowe PJ, Deniz Y, Lederer DJ, Hardin M. Dupilumab reduces oral corticosteroid use in patients with corticosteroid-dependent severe asthma: an analysis of the phase 3, open-label extension TRAVERSE trial. Chest. 2022. Epub 20220222. doi: 10.1016/j.chest.2022.01.071. PubMed PMID: 35217003.
32. Brands X, Uhel F, van Vught LA, Wiewel MA, Hoogendijk AJ, Lutter R, Schultz MJ, Scicluna BP, van der Poll T. Immune suppression is associated with enhanced systemic inflammatory, endothelial and procoagulant responses in critically ill patients. PLoS One. 2022;17(7):e0271637. Epub 20220725. doi: 10.1371/journal.pone.0271637. PubMed PMID: 35877767; PMCID: PMC9312372.
33. Turner NA, Mughal RS, Warburton P, O'Regan DJ, Ball SG, Porter KE. Mechanism of TNFalpha-induced IL-1alpha, IL-1beta and IL-6 expression in human cardiac fibroblasts: effects of statins and thiazolidinediones. Cardiovasc Res. 2007;76(1):81–90. Epub 20070612. doi: 10.1016/j.cardiores.2007.06.003. PubMed PMID: 17612514.

7 Aggregation and Solubility

Making Milk from Cheese

Remember me. [exit ghost]

William Shakespeare, *Hamlet*, Act 1 Scene 5, line 91[1]

OVERVIEW

1. Many different proteins and peptides can form stable aggregates as part of their functions, in spermatogenesis, blood clotting and forming fibers.
2. Acid, heat, salt, proteolysis and physical stress, all processes used in cooking, denature soluble proteins and induce aggregation. Aggregated proteins are difficult to resolubilize without adding chaotrophs, small molecules that degrade protein structure.
3. Many human proteins are quasi-stable; point mutations and post-translational changes, especially during aging, may lead to misfolding and formation of toxic aggregates.
4. Abnormal protein aggregation diseases can occur throughout the body, blocking blood and renal flows. Lambda light chain disease (LLCD) and other syndromes where antibodies or fragments block vessels are often related to B-cell disorders and cancers.
5. Protein plaques in the brain are hallmarks of many progressive neurodegenerative syndromes associated with aging, including Alzheimer's, Parkinson's, Lewy body, Huntington's diseases (AD, PD, LBD, HD, respectively), chronic traumatic encephalopathy (CTE), and ataxias. Except for CTE, related to severe head trauma, and mutations in aggregating proteins related to early development of AD, the causes of most slowly developing, progressive neuropathies are unclear.
6. Not all dementias are irreversible; some caused by nutritional deficiencies, infections or autoimmune responses to tumors may be treatable.
7. Clearance of aggregates may be limited by disruptions in autophagy or secretion pathways; these may also respond to specific treatments.

DOI: 10.1201/9781003333319-8

8. Adjuvant therapies may be used to treat symptoms, such as inflammation and excitability, or block specific proteases, but are typically not curative.
9. Therapeutics to prevent or clear aggregated protein plaques require long-term studies to demonstrate efficacy in slowly developing, chronic diseases. Differences in personal behavior, disease progression and environmental factors may affect results.

Figure 1.1 shows how easily methane can be converted to a potent toxin, just by changing one hydrogen atom to a halide. This chapter deals with another way to convert a protein into a toxin without changing its chemical formula, by forming aggregates. Simple chemical differences can affect the solubility and function of small molecules and polymers. Figure 7.1 gives a very simple example, based on Haworth's Noble Prize lecture (1937),[2] of just how important the orientation of two linked sugar rings to each other can be for their solubility and digestibility. If two glucose rings are linked in an **α** fashion, they form maltose, the basic component of starch, which is reasonably soluble in water. If the sugars are linked in a β linkage, with the inter-glucose bond reversed, they form cellobiose, the basic unit of cellulose. Cellulose is not water soluble and is not digested in humans (ruminants have additional stomachs to allow microbes to degrade hay and other food stuffs). Although the atoms of both molecules are essentially identical, the bond direction means that different enzymes, called amylases, for starch degradation, or cellulases for cellulosic materials, are needed to convert them to metabolizable glucose units. Similarly, changes in the amino acid orientations in proteins, such as proline flipping

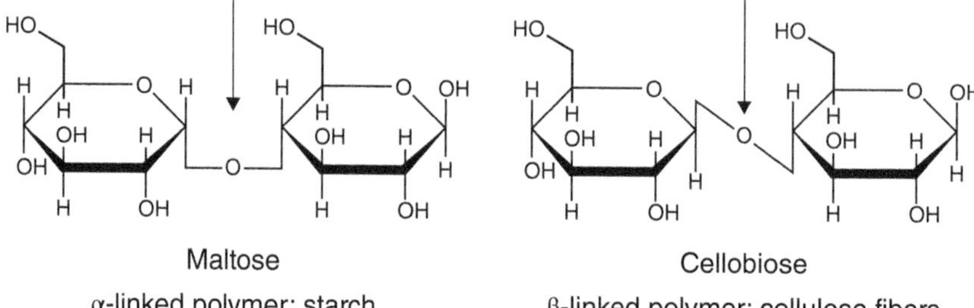

FIGURE 7.1 Bread and hay: Two glucose units have quite different solubility if linked in an **α** (maltose, left, polymers of which make up starch) or **β** (cellobiose, right, subunit of cellulose polymers) fashion. Protein conformations can change or aggregate to form insoluble plaques that resist proteolysis or autophagy.

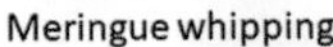

FIGURE 7.2 Protein aggregates in food can be generated through different processing methods. Traditional cooking applies heat, acid, salt, physical stress (whipping) and proteolysis to alter the solubility of egg and milk proteins to achieve quite different products.

from cis to trans, or cysteine residues forming alternative bonds, can lead to changes in their function. Alternatively, proteins can form aggregates, with positive or negative consequences.

Perhaps the easiest way to see proteins in the process of aggregating is during cooking. Careful use of heat, acid, salt, proteases (which cut the protein into smaller pieces) and physical stresses can denature and change the solubility of proteins, leading to desired textures and tastes (Figure 7.2). Heating will lead to the formation of insoluble protein, whether an egg is fried, boiled, mixed with hot rice or baked. One can whip parts of an egg alone (to form meringue from the whites) or with oil, salt and acid (to make mayonnaise from the yolk). Fish protein will turn opaque when lime juice and salt are added to make ceviche. Both acid and heat are used to form cheese from milk. The milk can be acidified naturally by encouraging the conversion of sugars to lactic acid by lactobacilli (whose growth is in turn favored by incubating the milk at around body temperature [37°C or about 99°F] to yield yogurt). The cheese that forms when the curds are removed from the whey contains primarily proteins, with some associated fat and residual sugars from the milk. A similar process is used to produce tofu from soy milk. Adding acids like vinegar or lemon juice can be used to make semihard cheeses (e.g., paneer and cottage cheese). To form hard cheese, proteases such as renin are added to enhance the denaturation process and resulting aggregation.

As any cook who has beaten egg whites for too long knows, controlling or reversing protein aggregation is very difficult. While heating a hard cheese will seem to "melt" it to a thick liquid (to make a proper fondue), this is due to melting fat while stirring to maintain an emulsion, stabilized by wine (which contains tartrate as emulsifier). Overheating will make cheese fondue a thinner liquid as the mass separates into sticky protein aggregates swimming in liquid fat and water (from that wine!). The fondue can "be saved" from this unappetizing state by adding some cornstarch dissolved in yet more wine.*

Regenerating milk from cheese is a difficult, perhaps unsolvable, problem.

AGGREGATION CAN CONTRIBUTE TO PROTEIN FUNCTION

Many proteins form stable aggregate structures, called amyloid. Amyloid formation, where proteins aggregate into self-organized structures,[3] is a necessary step in many biological processes.[4] Amino acid chains of proteins may change their orientation to each other to form insoluble sheets or plaques. Initially soluble proteins or peptides can form highly structured, functional aggregates, contributing to fertility and normal cell function. Even small peptides can form very stable β-sheet structures that have many different uses in tablet formulation and drug design. Silk forms from a highly repetitive protein, containing repeats of three small amino acids, serine, glycine and alanine, with a few scattered glutamates, aspartates and valines. The fibrils self-aggregate in an ordered fashion to form a β-sheet structure stabilized by hydrogen bonds.

Many other proteins form stable, functional aggregates. Amyloid fibrils may aid in clearing damaged sperm, thus facilitating reproduction.[5] Insulin in the pancreas is stored in acidic granules until release; other proteinaceous hormones are stored as reversible amyloids in secretory granules in a variety of glands. Pure insulin forms pharmaceutically unacceptable fibrils that do not easily redissolve to the active monomer needed to control glucose levels in the blood. Intermediate and long-acting forms of insulin are stabilized by the addition of Zn^{2+} ions, which enables the protein to form hexamers and controlled ordered structures that slowly dissolve.[6] Addition of specific amounts of zinc to recombinant insulin is one way to control the distribution and serum half-life of preparations, allowing diabetics to control their blood sugar with fewer injections per day.

More complicated proteins can aggregate as part of normal physiology. Fibrinogen in blood forms clots of fibrin after proteolysis (see below) during

* Authentic Swiss cheese fondue contains no cornstarch or flour, as these lead to a grainy texture, but nutmeg and significant amounts of Kirsch (very dry *eau de vie,* made from cherries by double distillation) are *de rigueur.*

wound healing. However, large clots (for example the dreaded deep vein thrombi (DVTs) that can form during a long period of inactivity) can block arteries, causing strokes and heart attacks if not degraded by anticoagulases.

BLOOD CLOTTING IS A YIN/YANG PROCESS

The proteolysis of fibrinogen to fibrin is the basis of clotting. Since blood clots are necessary for wound healing but can be deadly if produced in the circulatory system, the proteolysis leading to fibrin and formation of a stable clot is controlled by a complicated system of proteases, called "factors" that activate one another sequentially. The first step in clotting is accumulation of platelets at the site of injury. Thereafter, the clot is stabilized by a row of "serine" proteases (serine here refers to one of the amino acids in their active site, not where they cleave in the protein) that exist as inactive "zymogens" (think zombie enzymes) until they are activated during the "complement cascade" (Figure 7.3). Lacking one of the proteases in

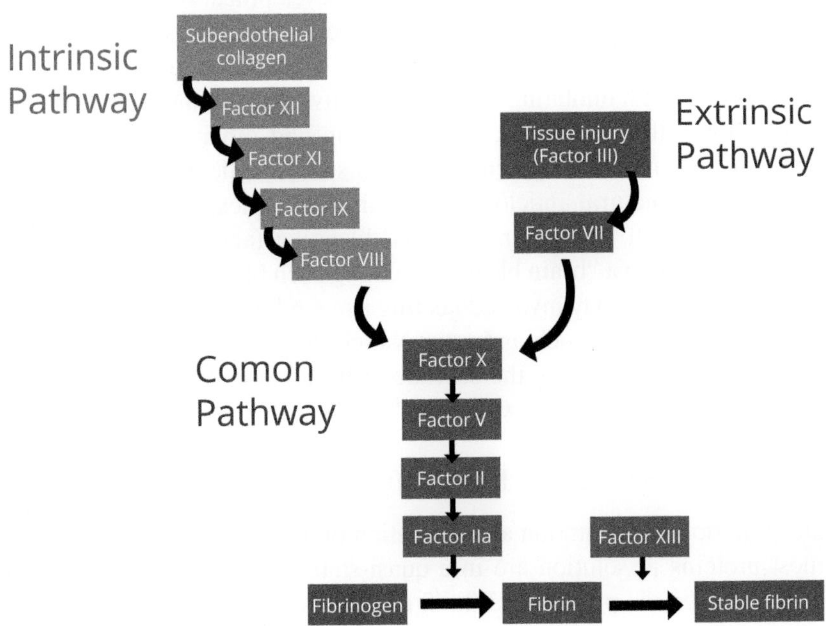

FIGURE 7.3 A complex cascade of blood factors controls the formation of blood clots from fibrin. The factors in the process are specific proteases that are activated by the protein above in the cascade once it begins.

the series, most famously Factor VIII or Factor IX,† causes different forms of hemophilia, characterized by the inability to stop bleeding if injured.[7]

Unfortunately, the clotting process can also be started by damaged sites or constrictions in the circulatory system, or when sitting for long periods on a plane. When large clots form, they can block blood vessels or travel to the brain or lungs, leading to dangerous ischemic strokes, difficulty breathing (pulmonary embolism), or potentially fatal heart attacks. Once a stroke has started, therapies, such as injection of the bacterial protease streptokinase (see the next chapter) or tissue plasminogen activator (tPA), are used to break up the fibrin protein aggregates in the clot. While streptokinase is much cheaper, tPA is usually preferred. It is more specific for fibrin and, as the recombinant form of the human protein, less likely to be antigenic. Either of these therapies may also be used for DVTs, depending on their size and location in the body. Compression socks and aspirin before long flights are recommended to prevent DVTs from forming.

Several different drugs are prescribed to reduce the potential for clotting. The most common therapies, low-dose aspirin, warfarin or heparin, interfere with fibrin aggregate formation. Other drugs bind specifically to one or several of the zymogen proteases in the clotting cascade and prevent their activation by other factors. The pharmaceutical industry has been involved for decades in designing and synthesizing specific protease inhibitors to control clotting. This work also aided in the rapid development of inhibitors of the viral proteases of Covid-19. Inhibitors of specific "secretases" involved in the proteolysis of amyloid precursor protein (see below) also reflect the decades of experience with inhibiting the protease activity of clotting factor enzymes.

Paradoxically, those prescribed warfarin and other anticlotting drugs are counseled not to eat green vegetables such as kale, which are the best sources of the natural anticlotting vitamin K complex. The fear is that the combination will lead to uncontrolled bleeding (and a different kind of stroke risk, due to a "brain bleed"). Treating such bleeds in a person taking clotting inhibitors may involve injecting a recombinant blood factor to start the clotting cascade, such as Factor Xa (clotting factors decrease in number as they proceed down the cascade, getting closer to the point of fibrin cleavage and aggregation), where the "a" stands for activated.

PROTEIN INSTABILITY LEADS TO AGGREGATION

Denaturation, precipitation and formation of aggregates can be sudden, as most proteins in solution are in a quasi-stable state. The fragile equilibrium between the protein and the fluid that surrounds it can be destroyed

---

† Lack of Factor IX function led to the hemophilia that complicated the lives of many of the descendants of Queen Victoria, most famously the Romanoffs and the Spanish royal family.

by small stresses and storage under suboptimal conditions.[8] This fragility is illustrated by the many situations in which one wishes to keep proteins in a native, active state and avoid aggregation. Just a small increase in the growth temperature (from 28–30°C to 37°C) of bacteria can induce recombinant proteins to aggregate, forming insoluble inclusion bodies.[9] While these insoluble protein granules can be completely unfolded with high concentrations of "chaotrophs", typically urea or guanidinium, it is very difficult to obtain completely native structures after removing these denaturants.[10] Producing proteins at low temperature[11] and in the presence of natural chaperones or foldases[12] can aid production of soluble interferons, cytokines and antibodies for pharmaceutical use. After isolation, proteins precipitate or self-hydrolyze in solution. Solutions of proteins used for therapy are "formulated" to include co-solvents (typically buffers to control pH) and stabilizing molecules, such as salts and other physiologically acceptable small molecules. A given additive, e.g. TMAO (trimethylamine oxide, a natural metabolite), may stabilize one protein[13] while causing precipitation in others.

Many considerations play a role in the production and handling of therapeutic proteins. In general, keeping proteins in solution can lead to degradation and aggregation problems, shortening their shelf life and complicating and increasing the price of cytokine preparations and specific antibodies. Solubility dictates whether a vaccine or protein-based therapy can be shipped as a lyophilized powder, which can be rapidly redissolved when sterile water is added (an ideal situation, especially if the dried preparation is not particularly heat labile). The cost and use of proteins are greatly increased if they must be frozen. Unfortunately, most proteins, including insulin, must be kept in solution with accurate temperature control throughout shipping and storage.

PROTEIN "AMYLOIDOSIS" CAN CAUSE SYSTEMIC DISEASE

Amyloidosis can cause diseases throughout the body and involve different types of proteins. Many proteins accumulate mutations or post-translational changes during aging that cause them to precipitate or be recognized as foreign by the immune system. Accumulation of modified proteins can lead to autoimmune diseases, resulting in altered mental and physical state and even dementias. In cases where the association of detected aggregates to the patient's symptoms is not clear, they may be called monoclonal gammopathy of uncertain significance (MGUS).[14]

Antibodies or their fragments are among the most common proteins involved in amyloidosis syndromes. In addition to interfering with the function of normal proteins they bind to, overproduced antibody fragments can accumulate in tissues throughout the body. They may block blood flow (leading to clots in veins and even arteries), pancreatic ducts (leading to diabetes) or the kidneys (resulting in reduced renal function). Lambda light chain disease (LLCD)[15] (also called AL amyloidosis) is characterized by the accumulation of the smaller (light) chain of antibodies. Accumulation of the light chains, which would normally be only produced in proportion to their heavy chain partners, may be diagnosed during examination for other conditions, such as colon problems.[16] LLCD is often a signal of serious B-cell disorders (B-cells produce antibodies), especially multiple myeloma. The same regime used to treat a B-cell cancer will usually reduce the antibodies and can help resolve LLCD (if tissue damage is not too advanced). For example, "CBD" cancer therapy, used for multiple myeloma and lymphoma, which combines cyclophosphamide, an alkylating agent used to treat leukemia, with bortezomib (Velcade© and other brand names), a proteasome inhibitor that stimulates apoptosis, and dexamethasone, an anti-inflammatory steroid that also interferes with the secretion of antibody by B-cells,[17] can correct the cancer and stop the formation of further antibody aggregates.

There are also chemotherapy-free treatments that can control the B-cells that play such a central role in LLCD, which may be better tolerated in frail, older individuals.[18] An 80-year-old woman presented with kidney disease and had high (69%) plasmacytes in her bone marrow, indicating a B-cell dysplasia.[19] The plasmacytes disappeared after treatment with lenalidomide (a thalidomide derivative used to treat multiple myeloma[20]) and methylprednisolone (an anti-inflammatory steroid that blocks antibody secretion by B-cells).

PROTEIN PLAQUES CHARACTERIZE MANY NEUROPATHIES OF AGING

PolyQ repeat diseases (HD and ataxias related to long stretches of repeat nucleotides and glutamine (Q) regions), which were discussed in Chapter 4, are just one group of diseases characterized by the accumulations of aggregated proteins. Table 7.1 lists some of the known neurodegenerative diseases associated with plaques of aggregated proteins in the brain, including the most common, AD, PD and amyotrophic lateral sclerosis (ALS). The proteins that aggregate may be aged or aberrant forms of normal proteins, or fragments that are not easily degraded.

In AD, the Aβ peptide, which shows neurotoxic activity and rapidly becomes insoluble in solution,[21] is cleaved off a larger protein, amyloid precursor protein (APP) (Figure 7.4). While the insoluble Aβ is 40–42 amino acids long, variable length forms with differing solubilities accumulate,

TABLE 7.1
Diseases Characterized by Protein Aggregation. While CTE (related to head trauma) and familial forms of the other diseases (related to genetic mutations) may have early onset, most of these diseases are typically associated with aging. AD, PD and ALS (in that order) are the three leading neurodegenerative diseases in aging populations.

Aggregation Disease	Aggregated Protein Involved	What It Affects
Neurodegenerative		
Alzheimer's disease (AD)	Aβ42 and other fragments of amyloid precursor protein (mostly extracellular); intracellular tangles of phospho-tau	Memory, accelerated brain shrinkage
Amyotrophic lateral sclerosis (ALS)	Accumulations of TDP-43, Cu/Zn superoxide dismutase 1, TAR DNA-binding protein of 43 kDa, and fused in sarcoma among others[22]	Selective and progressive loss of motoneurons in the spinal cord, brainstem and cerebral cortex
Chronic traumatic Encephalopathy (CTE)	Buildup of tau around blood vessels (found frequently in athletes and others with frequent head injury	Memory, attention deficit, confusion, personality changes, problems with balance and movement
Frontotemporal lobar degeneration, frontotemporal dementia (FTD)	Similar to ALS, TDP-43	Personality changes, may also affect movement and eventually memory problems similar to AD
Lewy body disease	α-synuclein accumulations	REM sleep disruptions, intellectual decline, dementia, hallucinations, eventually PD movement disorders
Parkinson's disease (PD)	α-synuclein and others	Movement, eventually brain dysfunction
Aggregates affecting other parts of the body:		
Alpha-1-antitrypsin (AAT) deficiency	Mutations in AAT (that also lead to 1–3% of COPD cases)	AAT aggregates in the liver and cannot inhibit elastase effects that cause tissue breakdown

(Continued)

TABLE 7.1 *(Continued)*
Diseases Characterized by Protein Aggregation

Aggregation Disease	Aggregated Protein Involved	What It Affects
Idiopathic pulmonary fibrosis	Fibrotic deposits of cells, multiple proteins[23]	Lungs, breathing capacity, eventually may require lung transplant
Inclusion body myositis (IBM)	Filament aggregates of different proteins accumulate in muscle	Muscles; progressive weakness and eventual wheelchair dependency[24]
Lambda light chain (LLCD) and related syndromes	Antibody light chains and fragments (particularly related to B-cell lymphomas and other cancers)	Protein accumulations block the tubules in the kidneys and heart and blood vessels

depending on the degree of cleavage by different proteases (chiefly β and γ secretases). But inhibitors of these enzymes have not proved useful for controlling AD,[25] perhaps because inhibiting single enzymes cannot stop progression of the degradation process once it has started. As Aβ aggregates, it becomes a toxic protein for neurons in cell culture, compromising cellular metabolism, starting from membrane instability (via perforations)

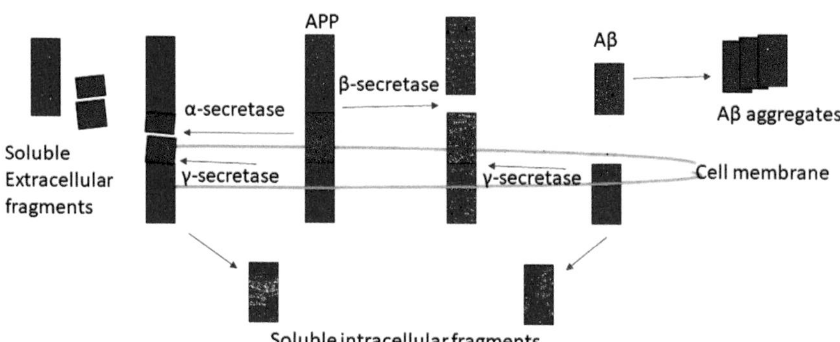

FIGURE 7.4 Cleavage of APP, amyloid protein precursor, can lead to short Aβ fragments that form aggregates and plaques in Alzheimer's disease. While processing through α- and γ- "secretases" leads to soluble and degradable fragments, cleavage through β- ("BACE") and γ- secretases leads to Aβ fragments that can form aggregates. Synthetically made Aβ also forms self-aggregates that are toxic to cultured neurons. BACE inhibitors have not been successful in clinical trials.

to signal amplification and recruitment of downstream signaling pathways that trigger degeneration. Aβ aggregates are toxic for neurons in cell culture,[21] causing them to shrivel and die. Simply exposing neural cells in culture to aggregated Aβ peptides (of any size) can result in substantial cell killing.[21]

Many question whether the visible aggregates of Aβ in the brain cells seen in AD autopsies are the basis of neurodegeneration, or just a way for the cell to wall off the aggregates. Some have suggested that the plaques in neurodegenerative diseases may help sequester dangerous protein oligomers, moderating their ability to kill cells. Further, there may be a cascade effect, in which Aβ aggregates cause other proteins, such as a much larger protein, *tau*, to aggregate as well.

ELIMINATING THE AGGREGATED PROTEIN MAY NOT BE THE WHOLE ANSWER

The plaques associated with many neuropathies are difficult to implicate as causative in disease process, unless removing them can clearly be shown to improve patient symptoms. Recently, antibodies that remove Aβ plaques also slowed cognitive loss when used at very early stages of AD characterized by mild to moderate cognitive impairment (called MCI).[26] It is still not clear that Aβ accumulation directly causes AD, as the antibody therapy also lowered the amount of a larger protein, phosphorylated-*tau* (see below), aggregates of which have also been associated with neurodegenerative diseases. There is an active debate on the possible conformational states or post-translational modifications of proteins[27] and the mechanisms that trigger neurodegeneration, ranging from reactive oxygen species (ROS) to altered synaptic plasticity (synapses are the connections between cells) via multiple pathways.[28]

Antibody therapies may not help in advanced AD cases, which are marked by brain shrinkage that is not easily reversible even if the plaques can be removed. There is thus active debate about when to administer treatment that might slow down disease progression, enhancing and extending the lives of patients. One encouraging result of recent serial brain scans of AD patients is that some white matter damage is reversible.

While the antibody results are promising, the AD field is riddled with failed drugs designed to inhibit proteins thought to contribute to the production of Aβ, such as inhibitors of secretases that cleave the peptide from APP, or accelerate its degradation.[29] As noted above, current success has come from treating early-stage patients. Another feature is that the patient pool was carefully selected (using advanced brain imaging) to have the plaques characteristic of AD.

IS IT REALLY AD?

One problem with early studies of cures for protein-aggregation-related diseases was that the patient population may have included those whose cognitive difficulties, dementia or depression, could have arisen from other causes (Figure 7.5, diverse causes of dementia). Infections (including Covid-19) can also trigger changes in cognition and cardiovascular aberrations. There are treatable causes that may lead to dementia symptoms in aging that may not involve aggregated proteins. Perhaps the easiest to treat are those due to vitamin deficiencies, such as a lack of vitamins B1 (thiamine), B3 (nicotinic acid), B6, B12 or nicotinic acid (seen first in pellagra).[30] Another reason may be "paraneoplastic neuronal syndromes" related to the formation of autoantibodies to tumor antigens.[31] Often the neurological symptoms are noted before the tumor has been diagnosed. The antibodies can invade the brain, attacking neurons and causing encephalitis,[32] ataxia and disorientation. Once the underlying antibodies are detected, the neurological symptoms may respond to treatments for immune disease, such as steroids. They may disappear completely after removal of the cancer. One now well-characterized syndrome is due to the formation of antibodies to the NMDA receptor on brain cells, which may be triggered by ovarian or cervical tumors.

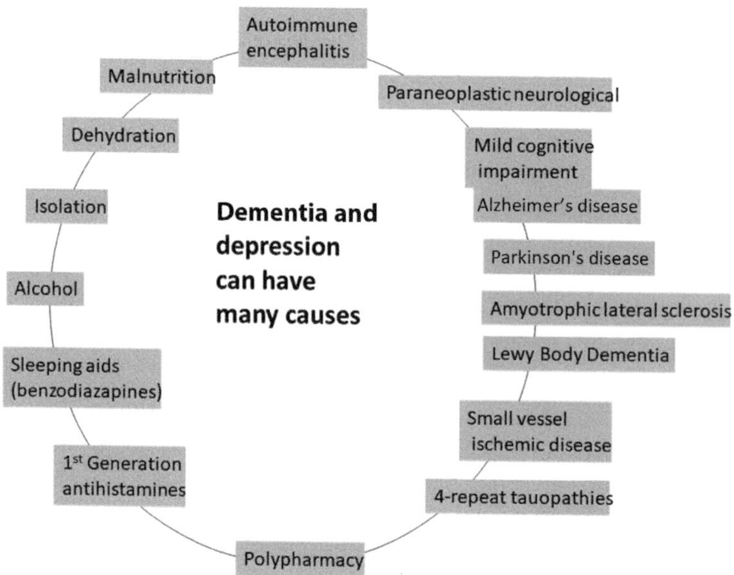

FIGURE 7.5 The psychiatric symptoms of dementia and depression can have many causes. Even dehydration or malnutrition in a frail elderly patient can lead to depression and cognitive symptoms of severe, progressive syndromes.

The success of recent studies has been to use more advanced, non-invasive brain imaging, using selective dyes in PET imaging, which give better ways to distinguish different forms of AD or PD and treatment options for co-morbidities.[33,34] MRI or PET scans of those with a clinical neurological AD diagnosis may reveal a much more complicated situation. Cerebrovascular disease and hippocampal sclerosis with aging[35] may be more common causes of dementia in those over 75.[36] Small (or micro-) vessel ischemic disease (SVID) may accompany AD or cause dementia on its own.[37] LBD, which can cause visual hallucinations and acting out, or frontotemporal dementias may be separate diseases or accompany AD and SVID.

AFTER MANY FAILURES, DRUGS TO TREAT NEURODEGENERATIVE DISEASES SHOW SOME SUCCESS

Aβ plaques (visualized with specific PET scan assays) can be removed by antibody treatment in those with MCI, but it may take years of follow-up studies ("Phase 4") to show this slows the long-term progress of the disease without excessive side effects.[30] A possible reason is that there have been so many failures in this area and that the true trigger for AD is not known; relying on eliminating the aggregated protein may be chasing only part of the disease. While memory problems are certainly of concern, they do not automatically mean progression to dementia and loss of function. There are other diseases that can cause similar symptoms to the diseases causing progressive functional and cognitive losses in Table 7.1. One syndrome that looks like AD in elderly people but has milder symptoms has a more complicated name: argyrophilic grain disease (AGD). Argyrophilic refers to aggregates of the *tau* protein that bind silver. The silver-coated protein granules show up as black grains in the brain tissue (where the protein causes the silver to change state in a process similar to what happens to table silver objects over time). The *tau* protein is phosphorylated in AGD to phospho-*tau (ptau)*. *Tau* fibrils also accumulate in PD, although the primary aggregating protein is α-synuclein.

Besides the time required for clinical trials, there are other major problems in determining the effectiveness of any treatment for chronic neurodegenerative disease. These include patient-to-patient variation in the time of progression from MCI to the later stages of AD. Another problem is certainly that preclinical models for testing any treatment are mice transformed with human proteins. It is difficult to accurately model a disease process that extends over decades in humans in an animal with a two-year life span.

CONCLUSIONS

Amyloid formation from initially soluble proteins can contribute to their function, in areas from molecular glues to spermatogenesis to blood clotting. On the other hand, formation of aggregates can contribute to diseases throughout the body. Many different neurodegenerative syndromes characterized by protein aggregates can now be distinguished by direct brain imaging. This has led to better controlled clinical trials of treatments proposed for AD and PD.

Agents that perturb aggregated protein plaques may be the basis of treatments for these diseases. The real treatments may come from drugs targeting the triggers that cause harmless, soluble proteins to become toxic aggregates.

REFERENCES

1. Shakespeare W. The Complete Works of William Shakespeare. London, England: Cambridge University Press, Octopus Books Ltd.; 1980.
2. Haworth WN. The Structure of Carbohydrates and of Vitamin C. Stockholm, Sweden: Nobelprize.org; 1937. Available from: https://www.nobelprize.org/uploads/2018/06/haworth-lecture.pdf.
3. Kokotidou C, Tamamis P, Mitraki A. Chapter 7 Amyloid-Like Peptide Aggregates. Peptide-based Biomaterials. The Royal Society of Chemistry; 2021. p. 217–68. https://doi.org/10.1039/9781839161148
4. Otzen D, Riek R. Functional amyloids. Cold Spring Harb Perspect Biol. 2019;11(12). Epub 20191202. doi: 10.1101/cshperspect.a033860. PubMed PMID: 31088827; PMCID: PMC6886451.
5. Roan NR, Sandi-Monroy N, Kohgadai N, Usmani SM, Hamil KG, Neidleman J, Montano M, Standker L, Rocker A, Cavrois M, Rosen J, Marson K, Smith JF, Pilcher CD, Gagsteiger F, Sakk O, O'Rand M, Lishko PV, Kirchhoff F, Munch J, Greene WC. Semen amyloids participate in spermatozoa selection and clearance. Elife. 2017;6. doi: 10.7554/eLife.24888. PubMed PMID: 28653619; PMCID: PMC5487211.
6. Frankaer CG, Sonderby P, Bang MB, Mateiu RV, Groenning M, Bukrinski J, Harris P. Insulin fibrillation: The influence and coordination of Zn(2). J Struct Biol. 2017;199(1):27–38. Epub 20170518. doi: 10.1016/j.jsb.2017.05.006. PubMed PMID: 28527712.
7. Lannoy N, Hermans C. The 'royal disease' – haemophilia A or B? A haematological mystery is finally solved. Haemophilia. 2010;16(6):843–7. doi: 10.1111/j.1365-2516.2010.02327.x. PubMed PMID: 20557352.
8. Schein CH. Solubility as a function of protein structure and solvent components. Bio/Technology. 1990;8(4):308–17. doi: 10.1038/nbt0490-308.
9. Schein CH. Optimizing protein folding to the native state in bacteria. Curr Opin Biotechnol. 1991;2(5):746–50. doi: 10.1016/0958-1669(91)90046-8. PubMed PMID: 1367729.

10. Schein CH. Protein Aggregation and Precipitation, Measurement and Control. 2010: John Wiley & Sons online library. https://doi.org/10.1002/9780470054581.eib052
11. Schein CH. A cool way to make proteins. Nat Biotechnol. 2004;22(7):826–7. doi: 10.1038/nbt0704-826. PubMed PMID: 15229543.
12. Baker WS, Negi S, Braun W, Schein CH. Producing physicochemical property consensus alphavirus protein antigens for broad spectrum vaccine design. Antiviral Res. 2020;182:104905. Epub 20200812. doi: 10.1016/j.antiviral.2020.104905. PubMed PMID: 32800880.
13. Schein CH, Oezguen N, Volk DE, Garimella R, Paul A, Braun W. NMR structure of the viral peptide linked to the genome (VPg) of poliovirus. Peptides. 2006;27(7):1676–84. Epub 20060315. doi: 10.1016/j.peptides.2006.01.018. PubMed PMID: 16540201; PMCID: PMC1629084.
14. Ueno M, Kobayashi S, Asakawa S, Arai S, Nagura M, Yamazaki O, Tamura Y, Ohashi R, Shibata S, Fujigaki Y. Emergence of proteinase 3-antineutrophil cytoplasmic antibody-associated glomerulonephritis with mesangial immune deposition during the clinical course of IgG lambda monoclonal gammopathy of uncertain significance. CEN Case Rep. 2022. Epub 20220414. doi: 10.1007/s13730-022-00703-4. PubMed PMID: 35420387.
15. Berghaus N, Schreiner S, Granzow M, Muller-Tidow C, Hegenbart U, Schonland SO, Huhn S. Analysis of the complete lambda light chain germline usage in patients with AL amyloidosis and dominant heart or kidney involvement. PLoS One. 2022;17(2):e0264407. Epub 20220225. doi: 10.1371/journal.pone.0264407. PubMed PMID: 35213605; PMCID: PMC8880859.
16. Alnimer L, Zakaria A, Patel J, Samhouri Y, Ahsan S, Goldman L, Sorser S. A case of systemic AL amyloidosis diagnosed by screening colonoscopy. Case Rep Gastrointest Med. 2022;2022:5562281. Epub 20220420. doi: 10.1155/2022/5562281. PubMed PMID: 35497055; PMCID: PMC9046006.
17. Porcari A, Pagura L, Rossi M, Porrazzo M, Dore F, Bussani R, Merlo M, Sinagra G. Light-chain cardiac amyloidosis: a case report of extraordinary sustained pathological response to cyclophosphamide, bortezomib, and dexamethasone combined therapy. Eur Heart J Case Rep. 2022;6(4):ytac130. Epub 20220322. doi: 10.1093/ehjcr/ytac130. PubMed PMID: 35652085; PMCID: PMC9149786.
18. Schein CH. Distinguishing curable from progressive dementias for defining cancer care options. Cancers. 2023;15(4):1055. PubMed PMID: doi:10.3390/cancers15041055.
19. Mima A, Nagahara D, Tansho K. Successful treatment of nephrotic syndrome induced by lambda light chain deposition disease using lenalidomide: A case report and review of the literature. Clin Nephrol. 2018;89(6):461–8. doi: 10.5414/CN109342. PubMed PMID: 29393843.
20. Schein CH. Repurposing approved drugs on the pathway to novel therapies. Med Res Rev. 2020;40(2):586–605. Epub 20190820. doi: 10.1002/med.21627. PubMed PMID: 31432544; PMCID: PMC7018532.
21. Chen D, Martin ZS, Soto C, Schein CH. Computational selection of inhibitors of Abeta aggregation and neuronal toxicity. Bioorg Med Chem. 2009;17(14):5189–97. Epub 20090527. doi: 10.1016/j.bmc.2009.05.047. PubMed PMID: 19540126; PMCID: PMC2743868.

22 Gosset P, Camu W, Raoul C, Mezghrani A. Prionoids in amyotrophic lateral sclerosis. Brain Commun. 2022;4(3):fcac145. Epub 20220609. doi: 10.1093/braincomms/fcac145. PubMed PMID: 35783556; PMCID: PMC9242622.
23 Calabrese F, Lunardi F, Tauro V, Pezzuto F, Fortarezza F, Vedovelli L, Faccioli E, Balestro E, Schiavon M, Esposito G, Vuljan SE, Giraudo C, Gregori D, Rea F, Spagnolo P. RNA sequencing of epithelial cell/fibroblastic foci sandwich in idiopathic pulmonary fibrosis: new insights on the signaling pathway. Int J Mol Sci. 2022;23(6). Epub 20220319. doi: 10.3390/ijms23063323. PubMed PMID: 35328744; PMCID: PMC8954546.
24 McLeish E, Slater N, Sooda A, Wilson A, Coudert JD, Lloyd TE, Needham M. Inclusion body myositis: the interplay between ageing, muscle degeneration and autoimmunity. Best Pract Res Clin Rheumatol. 2022:101761. Epub 20220624. doi: 10.1016/j.berh.2022.101761. PubMed PMID: 35760741.
25. Pardo-Moreno T, Gonzalez-Acedo A, Rivas-Dominguez A, Garcia-Morales V, Garcia-Cozar FJ, Ramos-Rodriguez JJ, Melguizo-Rodriguez L. Therapeutic approach to Alzheimer's disease: current treatments and new perspectives. Pharmaceutics. 2022;14(6). Epub 20220524. doi: 10.3390/pharmaceutics14061117. PubMed PMID: 35745693.
26. van Dyck CH, Swanson CJ, Aisen P, Bateman RJ, Chen C, Gee M, Kanekiyo M, Li D, Reyderman L, Cohen S, Froelich L, Katayama S, Sabbagh M, Vellas B, Watson D, Dhadda S, Irizarry M, Kramer LD, Iwatsubo T. Lecanemab in early Alzheimer's disease. N Engl J Med. 2023;388(1):9–21. Epub 20221129. doi: 10.1056/NEJMoa2212948. PubMed PMID: 36449413.
27. Marcatti M, Fracassi A, Montalbano M, Natarajan C, Krishnan B, Kayed R, Taglialatela G. Abeta/tau oligomer interplay at human synapses supports shifting therapeutic targets for Alzheimer's disease. Cell Mol Life Sci. 2022;79(4):222. Epub 20220404. doi: 10.1007/s00018-022-04255-9. PubMed PMID: 35377002; PMCID: PMC8979934.
28. Cummings J, Lee G, Nahed P, Kambar M, Zhong K, Fonseca J, Taghva K. Alzheimer's disease drug development pipeline: 2022. Alzheimers Dement (N Y). 2022;8(1):e12295. Epub 20220504. doi: 10.1002/trc2.12295. PubMed PMID: 35516416; PMCID: PMC9066743.
29. Schneider LS. Chapter 19 – Clinical issues in Alzheimer drug development. In: Wolfe MS, editor. Developing Therapeutics for Alzheimer's Disease. Boston: Academic Press; 2016. p. 503–21.
30. Preskin SM. A Molecule Away from Madness: Tales of the Hijacked Brain. New York: W. W. Norton & Company; 2022. 240 p.
31. Jarius S, Brauninger S, Chung HY, Geis C, Haas J, Komorowski L, Wildemann B, Roth C. Inositol 1,4,5-trisphosphate receptor type 1 autoantibody (ITPR1-IgG/anti-Sj)-associated autoimmune cerebellar ataxia, encephalitis and peripheral neuropathy: review of the literature. J Neuroinflammation. 2022;19(1):196. Epub 20220730. doi: 10.1186/s12974-022-02545-4. PubMed PMID: 35907972; PMCID: PMC9338677.
32. Behrman S, Lennox B. Autoimmune encephalitis in the elderly: who to test and what to test for. Evid Based Ment Health. 2019;22(4):172–6. Epub 20190919. doi: 10.1136/ebmental-2019-300110. PubMed PMID: 31537612.

33. Ali DG, Bahrani AA, Barber JM, El Khouli RH, Gold BT, Harp JP, Jiang Y, Wilcock DM, Jicha GA. Amyloid-PET levels in the precuneus and posterior cingulate cortices are associated with executive function scores in preclinical Alzheimer's disease prior to overt global amyloid positivity. J Alzheimers Dis. 2022;88(3):1127–35. doi: 10.3233/JAD-220294. PubMed PMID: 35754276.
34. Staffaroni AM, Elahi FM, McDermott D, Marton K, Karageorgiou E, Sacco S, Paoletti M, Caverzasi E, Hess CP, Rosen HJ, Geschwind MD. Neuroimaging in dementia. Semin Neurol. 2017;37(5):510–37. Epub 20171205. doi: 10.1055/s-0037-1608808. PubMed PMID: 29207412; PMCID: PMC5823524.
35. Nelson PT, et al., Hippocampal sclerosis of aging, a prevalent and high-morbidity brain disease. Acta Neuropathol. 2013;126(2):161–77. Epub 20130718. doi: 10.1007/s00401-013-1154-1. PubMed PMID: 23864344; PMCID: PMC3889169.
36. Nelson PT, et al., Correlation of Alzheimer disease neuropathologic changes with cognitive status: a review of the literature. J Neuropathol Exp Neurol. 2012;71(5):362–81. doi: 10.1097/NEN.0b013e31825018f7. PubMed PMID: 22487856; PMCID: PMC3560290.
37. Clancy U, Gilmartin D, Jochems ACC, Knox L, Doubal FN, Wardlaw JM. Neuropsychiatric symptoms associated with cerebral small vessel disease: a systematic review and meta-analysis. Lancet Psychiatry. 2021;8(3):225–36. Epub 20210201. doi: 10.1016/S2215-0366(20)30431-4. PubMed PMID: 33539776.

8 Converting Bacterial Toxins to Human Therapeutics

They [the alien invaders] were undone, destroyed, after all of man's weapons and devices had failed, by the tiniest creatures that God in his wisdom put upon this earth. By the toll of a billion deaths, man had earned his immunity, his right to survive among this planet's infinite organisms. And that right is ours against all challenges. For neither do men live nor die in vain.

HG Wells, *The War of the Worlds*

OVERVIEW

1. Bacterial toxins evolved in mixed microbial communities to increase the survival of producing cells.
2. Many were derived from intracellular enzymes, including proteases, nucleases, lipases and adenylyl cyclases.
3. Their intracellular toxicity in the producing organism is usually tightly controlled by specific inhibitor binding or encapsulation in missile or syringe-like injection systems.
4. The coding sequences are frequently found on plasmids that can be passed even to different bacterial species.
5. Very different pathogens produce similar toxins when invading mammalian hosts.
6. Staphylococcal and streptococcal toxins are examples of "superantigens" that can induce immune system overreaction and cytokine release contributing to septic or "toxic shock syndrome".
7. Many virulence factors, their inhibitors and accessories are conditionally toxic proteins that can be exploited as research tools, analgesics, and for immune evasion and anticancer therapeutics.
8. Inactivated extracellular toxins are the basis of safe and inexpensive vaccines against bacteria that cause diphtheria, tetanus and pertussis, among others.

DOI: 10.1201/9781003333319-9

9. Even the deadliest toxins, including the neurotoxin of *Clostridium botulinum,* have found use in research, medicine and cosmetics.

INTRODUCTION

Pathogenic organisms are released whenever contaminated water seeps into our lives, adding to the agony of populations suffering from wars and natural disasters. They take advantage of those times when it is most difficult to provide hospitalization, clean water and good nursing care to attack whole populations, feasting on the blood and tissues of young and old. Although today we have better methods for rehydrating patients after the fluid loss due to infection with toxin-producing shigella or cholera strains, antibiotics typically do not help treat diarrhea and can even make the situation worse. The pathway to cure may indeed be through specific inhibitors, such as the small molecules that inhibit different toxins. New antibacterials and vaccines may conversely lie in manipulating the very toxins responsible for the deaths of so many animals and humans.

IT'S A MOLD KILLS BACTERIA AND PHAGES EAT ALL WORLD

Even a small amount of pond water can contain more individual microorganisms than all the mammals who ever walked the earth. Louis Pasteur, the great microbiologist who pioneered the preservation of milk by heating, published a series of papers in the 19th century showing that one bacterial strain could interfere with the growth of another. He proved that the interfering substance was probably a small protein, what we now call a toxin, as a cell-free filtrate of bacteria inhibited another strain as well as when whole cells were mixed. Bacteria used phages (also referred to as bacterial viruses) as a way to envelope toxins produced within their own cytoplasm. Once the phage was released, it could attack and enter other cells through its lysin proteins and release the killer toxins into the enemy cytoplasm. The phages released by infected bacteria could go on to infect other strains, carrying with them the genes for many other virulence factors.

Pasteur also showed isolated supernatants could kill mammalian cells. While, as discussed in Chapter 6, Coley sought to use the toxins produced by *Streptococci* to cure cancer, Pasteur dreamed of harnessing the byproducts of microorganisms to kill other pathogenic bacteria. In the great scheme of things, he reasoned that products produced by viruses should kill bacteria or eukaryotic cells, while those produced by molds should kill bacteria and prevent viruses from invading human cells. Indeed, Alexander Fleming would discover one mold, *Penicillium*, whose extracellular fluid contained the basis for the wide spectrum antibiotics that now protect us from many different bacteria.

Probiotic studies have indicated that encouraging nonpathogenic bacteria can protect against overgrowth of more dangerous ones, but probiotics cannot stop an infection once it reaches the blood stream (referred to as sepsis). However, treatment with bacterial phages, or isolated lysins they produce, that specifically destroy bacterial cell walls, is becoming a reality. Phage preparations have been used sporadically for over 100 years, in food treatment and against cholera and animal diseases. Human therapies (with or without continuing antibiotics) have received compassionate approvals by the FDA in patients with multidrug-resistant (MDR) infection, and treatment with isolated lysins is also being tested. Exebacase, a lysin against MDR *Staphylococcus aureus* (MRSA), gave unconvincing results in phase 3 clinical trials compared to "state of care" antibiotic treatment, to overcome antibiotic-resistant and biofilm-forming infections. However, it may still have a future in severly ill patients with antibiotic sensitivities. Newly engineered versions of lysins are being developed against gram-negative organisms as well.

BACTERIAL SPECIES COMPETITION: THE ONGOING WAR

We think of pathogenic bacteria as our enemy, enhancing their abilities to damage people while supporting their own survival. However, the human body was not their initial battle zone. Bacteria, and specific phages that could replicate within them, were killing off other microbes eons before the advent of mammals. The average bacterium lives in an environment full of predatory organisms. Toxins, their inhibitors and sophisticated plasmid and phage-encoded systems for escaping their producers' cell walls and penetrating the cytoplasms of competitors were all developed as offensive weapons in ongoing microbial wars. Small peptides and many larger toxins have sequence, structure or functional similarity to intracellular effectors and enzymes. Most toxins probably originated from bacterial enzymes by a process of directed evolution. As bacterial cells grow and evolve very quickly, mutants of these would be selected if they contributed to bacterial survival. Chance mutations in proteins that gave some survival benefits were retained and propagated by the descendants of the most successful cells. Developed in one organism, genes from successful toxin producers passed even to cells of other species.

Multicellular organisms and humans may be primary or secondary targets of these weapons. The skin, mouth, nose, lungs and digestive system contain diverse mixtures of bacteria, viruses and fungi, called our "microbiome". When the microbiome functions as it should, it protects

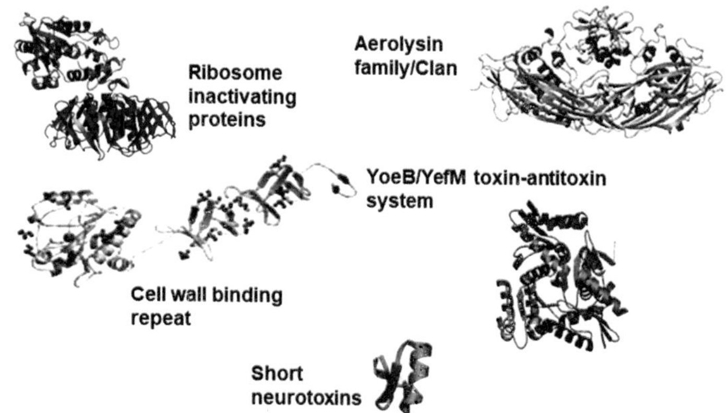

FIGURE 8.1 Examples of the diverse structures of toxin families produced by bacteria. These include ribosome inactivating toxins (1TFM.pdb), cell wall binding repeat (green strands) of autolysin LytA from *Streptococcus pneumoniae* TIGR4 (4X36.pdb); short neurotoxins (1COE.pdb); complex of YoeB/YefM from *Escherichia coli* (2A6Q.pdb); *Aeromonas* aerolysin pore forming units (5 JZH.pdb).

the sterile (i.e., free of measurable microbes) core of the body, aids digestion and discourages the growth of pathogens. However, occasionally a microbe will overwhelm these controls, invade normally sterile areas and cause widespread tissue destruction, aided by its array of toxins. Very common human bacterial pathogens used in screens for novel antibiotics are often referred to as "ESKAPE" (for *Enterococcus, Staphylococcus, Klebsiella, Acinetobacter, Pseudomonas, Enterobacter*) species. But many other pathogenic bacteria, including *Shigella* (basically a toxin-producing *Escherichia coli*), *Bacillus anthracis*, *Vibrio* and *Clostridia* species, produce virulence factors that aid in human infections. Figure 8.1 shows a few examples to illustrate how diverse bacterial toxin protein structures can be.

A few of the reasons bacteria produce toxins include:

1. preventing the growth of competing microbes
2. creating an ideal growth environment for themselves
3. allowing them to penetrate membranes or other structures
4. evading or stimulating the immune system of their host
5. limiting their own growth under conditions where active growth is not possible

Since toxins are produced for all these reasons, it is obvious why they have so many different activities. As with the proteins discussed in previous chapters, knowing the mechanism of activity allows toxins to be exploited for positive uses.

Toxins Are Key Tools to Attack Other Microbes, Even in Our Microbiome

As described in the introduction, many extracellular toxins inhibit the growth of competing microbes. Some of the most common bacterial toxins are nucleases that can degrade DNA or RNA, proteases, glycosidases, muramidases and lipases that break down the polysaccharides and other components of the bacterial cell wall and disrupt metabolism. Pore-forming and genotoxic virulence factors cause disruption of cell membranes and transport systems and induce DNA damage. Many of these virulence factors are carried into cells by phages, which also encode lysin proteins (Plys). The Plys genes target a particularly well-conserved site in the bacterial cell wall, facilitating the transfer of toxin genes and virulence factors from one organism to another, rendering those that survive the attack toxin producers. MDR bacteria may produce a defensive arsenal of immunity proteins and export pumps[1] to remove antibiotics from their cytoplasms.

In a well-balanced microbiome, diverse microbes can exist together in equilibrium. Tampering with this system, for example by treating with wide-spectrum oral antibiotics that alter the balance of antibiotic sensitive, MDR organisms in the human gut, may allow pathogens to take over. One of the most well-known examples of this is infection with the MDR organism, *Clostridium difficile* (Cdiff). Although Cdiff is an occupant of normal microbiomes, and indeed found on outdoor surfaces and shoe soles throughout the world,[2] it can overgrow other helpful organisms in the intestine, especially in those previously treated with antibiotics. Infections can spread rapidly in hospitals and rest homes. Aided by virulence factors (including the large toxins TCD-A and -B[3]), Cdiff forms deep crypts in the intestine, causing bloody diarrhea. Transplanting healthy fecal microbiota, thereby reintroducing diverse microbes to restore a working microbiome, has had some success as a Cdiff treatment, although other ways to inhibit this organism are being sought.

Monotypes in controlled environments may be much more susceptible to killer bacteria

Conversely, when we select organisms that do not produce toxins to grow in fermenters, to produce single-cell protein (SCP), fungal mats to make artificial meats or recombinant proteins, we also remove the organisms' ability to fight off inadvertently introduced wild strains. Cultivated strains sacrifice the genes for "immunity proteins" that allow them to resist the toxins of other microorganisms and accelerate their growth rate. Even protein production characteristics on plasmids selected for initially with antibiotic resistance drop during extended culture if they come with a growth disadvantage.[4] Stripped of all their defensive strategies, the selected monocultures do not stand a chance in the ongoing microbial wars. Sheltered, well-fed monocultures in a human-defined growth medium will be overgrown. A desired culture will "lose the war" to a wild type, toxin-producing contaminant organism or any faster growing mutant. Cultures grown for extended times will lose the plasmids or gene segments[5] encoding a desired recombinant protein.

Symbiotic bacteria living on plants or animals grown in a non-natural environment may also lose their defenses against an armed invader, as seen in infections on plants grown in hydroponic culture, or ones that overgrow fish farms, causing a "dirt taste" in cultured tilapia.[6] A bacterium isolated from one hydroponic farm was able to kill at least six different bacteria on contact, possibly through toxins injected with its evolved Type 6 secretion system (T6SS, see below).[7]

Exploiting the antimicrobial properties of bacterial toxins

Diabetic foot ulcers and infections of wounds and breaks that resist all antibiotic treatments have shown the pressing need for new treatments for MDR infections. A developing therapy takes advantage of the lysins (Plys) that allow phages to enter and lyse bacteria. There have been several successful attempts to take advantage of the spectrum of specific antibacterial factors that allow phage to cause lysis of MDR pathogens, such as MRSA and strains of Pseudomonas, Aeromonas baumannii, Enterococci and Cdiff. Penetrating and destroying specific MDR bacteria involves selecting phages (among the thousands known not to contain genes for toxins active against mammalian cells) or lysins that will specifically kill only pathogenic bacteria. Although there have been miraculous "phage cures", currently there are no clear rules that can help select which phage will kill a given pathogen.

Bacillus species secrete a large variety of bacteriocins, ranging in size from small peptides to whole proteins of more than 30 kD. One of these is the well-known antibacterial bacitracin, a small cyclic peptide found in many antibiotic creams, which must be topically used as it can cause kidney damage if used internally. Other bacteriocins are used in agriculture, as antimicrobials in fish farming and against the pathogens that grow on tomatoes and other plants. Some bacteriocins can even inhibit MRSA strains, opening the way for novel treatments in humans and animals.[8]

Toxins Modulate the Environment to Allow Their Producers to Exist

Mammals have their own defensive weapons against infection, part of the miraculous healing power of the human body. If a bacterium or virus enters a wound and tries to survive in our sera, it will usually be eliminated by a rapid, multi-pronged attack. The first level is circulating proteolytic enzymes that rapidly degrade any foreign proteins. A second line of defense is macrophages, which can wrap around the invaders and eventually destroy them. A third defensive line is specific antibodies, that bind to targets on the surface of the bacteria and direct complement proteases to them. Cytotoxic and "Natural killer" T-cells can also make a direct attack on the invaders. Bacteria without specific defenses against these mammalian weapons will rapidly lyse and be eliminated.

But especially in diabetics or immunosuppressed individuals who have weaker defenses, the toxins of successful pathogens such as MRSA, Group A *Streptococcus* (GAS)[9] or the water-born *Vibrio vulnificus* or *Aeromonas*[10] can break through all these barriers. The GAS-SpeB protease directly activates IL-1β outside of its normal pathway (see Chapter 6, Storming Cytokines) to start an inflammatory response (the red throat characteristic of GAS infections) that *enhances* the pathogen's growth. MRSA, on the other hand, wraps itself in a protective biofilm and secretes virulence factors that subdue most of the innate immune response and cytokine release.[11]

Some of the most dangerous toxins target mammalian cell membranes, inducing lysis directly, or have accessory proteins that enable them to cross the intestinal barrier. A Shiga toxin (Stx) producing strain of *E. coli* (O104:H4), tied to 810 cases and 39 deaths of hemolytic uremic syndrome,[12] produced adhesion factors that enhanced its ability to bind to intestinal epithelial cells as well as an arsenal of β-lactamases (which inactivate penicillin-based antibiotics). Human-specific, protease virulence factors cleave collagen, which makes up as much as 30% of the body, or other membrane proteins. Other toxins can depolymerize the actin filaments that underlie the surface of mammalian cells, leading to membrane instability and lysis.[13]

In the most extreme case, "necrotizing fasciitis", protease and collagenase toxins (for example of sea-borne *Vibrio* species) eat their way through soft tissues. Other virulence factors enhance vascular permeability, allowing the bacteria to spread subcutaneously and along the connective collagen fibers (fascia). If not caught quickly, and eliminated by infused antibiotics and debridement, whole limbs may need to be amputated and the patient may succumb to septicemia. Streptococcal infections (typically group A and B, GAS and GBS), referred to as puerperal fever or maternal sepsis, were a leading cause of death for both mother and child before the advent of antibiotics. Even with antibiotic treatment, GAS infections of premature newborns can lead to potentially fatal necrotizing enterocolitis.

Many pathogens produce a similar toxin. When the bacteria that cause anthrax (*B. anthracis*), whooping cough (*Bordetella pertussis*) or plague (*Yersinia pestis*) infect humans, they release toxins into the body, which cause a massive increase in cAMP levels in tissues or the intestines. As cAMP levels rise, they interfere with host metabolism, hence working to the invader's advantage. Elevated cAMP levels allow more water into the intestine, leading to painful edema and diarrhea. Other bacteria, including *Vibrio cholerae*, have even more pernicious ways to produce high levels of cAMP. They secrete toxins that modify the enzymes of human cells, inducing their host to secrete more cAMP into bodily fluids. One of the extracellular toxins of anthrax (Figures 8.2 and 8.3), edema factor (EF), once inside the mammalian cells, converts ATP, a high-energy molecule that fuels cells, to cAMP.[14–16] For this reason, the intracellular enzymes that convert ATP to cAMP, called adenylyl cyclases, are usually controlled by binding to another protein in the cell, or by activating them with another molecule. High concentrations of cAMP are a signal to the cell to slow its replication, as ATP energy levels are low. Most successful pathogenic bacteria generate toxins that increase the cAMP levels in tissues, perhaps to slow host cell replication (although the survival of persister bacteria [see below] suggests there may be yet another reason).

Just why elevating cAMP is advantageous for the invading bacteria is not known. However, the toxic effects on its host are clear. Elevated cAMP plays havoc with the proteins that control transport of sodium and

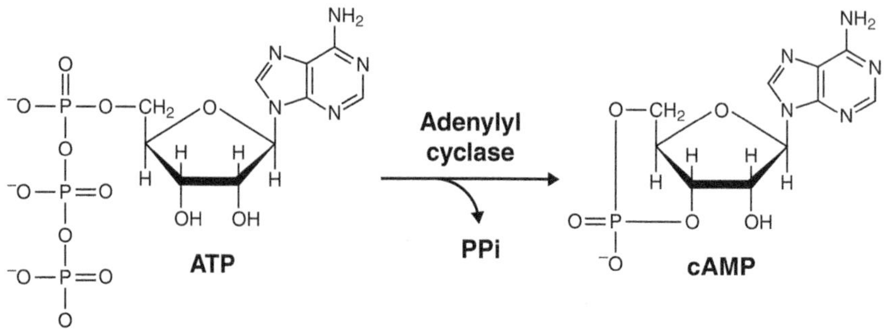

FIGURE 8.2 **Adenylyl cyclase toxins** convert ATP (left), the energy source in cells, to cyclic AMP (cAMP, right, the arrow indicates the circle formed after elimination of diphosphate). High cAMP levels in tissues contribute to edema and diarrhea in infectious diseases such as cholera and anthrax.

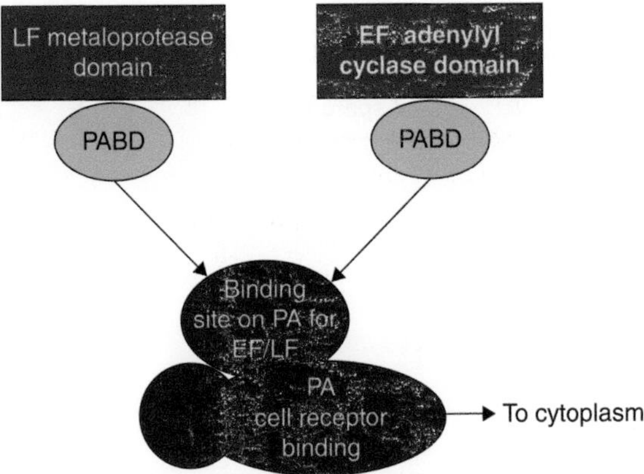

FIGURE 8.3 One pathway for two invading protein toxins. Toxins of anthrax enter cells by a channel made by the "protective antigen" (PA). Lethal factor, a protease, and edema factor, an adenylyl cyclase, both are preproteins, with a similar protective antigen binding domain (PABD) at their N-termini.[31]

potassium into and out of cells. High levels in the gut stimulate influx of fluid, causing edema. *V. cholerae* produces more than 20 different virulence factors, one of which stimulates cellular adenylyl cyclases, thereby elevating cAMP levels 100×. Left untreated, diarrhea associated with cholera infection can turn a human into a "cold gray corpse" within hours. Other pathogens secrete toxins that can function in the serum or within cells to enhance cAMP. To illustrate just how important cAMP-producing mechanisms are to bacteria, the intracellular pathogen *Mycobacterium tuberculosis* produces 12 different enzymes that generate cAMP!

Treatment with lethal toxins with anticlotting, antitumor and immune regulatory activities can even replace surgery

The same activities that contribute to pathogenesis have led to many uses of bacterial toxins. The plasminogen-activating toxin streptokinase, produced by many streptococci, is used to break up clots that cause strokes and block blood vessels throughout the body. Some of the bacteriocins produced by *Bacillus* species are pore forming and have anticancer properties. Many other pathogens produce toxins potentially useful in cancer therapy.[17]

Collagen is a primary protein component of all animals, essential for the flexibility of joints. However, in Dupuytren's disease, it is overproduced; nodules and extended fibers form, eventually leading to inability to move fingers and hands. Injection of a collagenase toxin secreted by *Clostridium histolyticum*, which can be done in an office visit, gives equivalent results to surgery in terms of restoring hand function, at a much lower cost.[18]

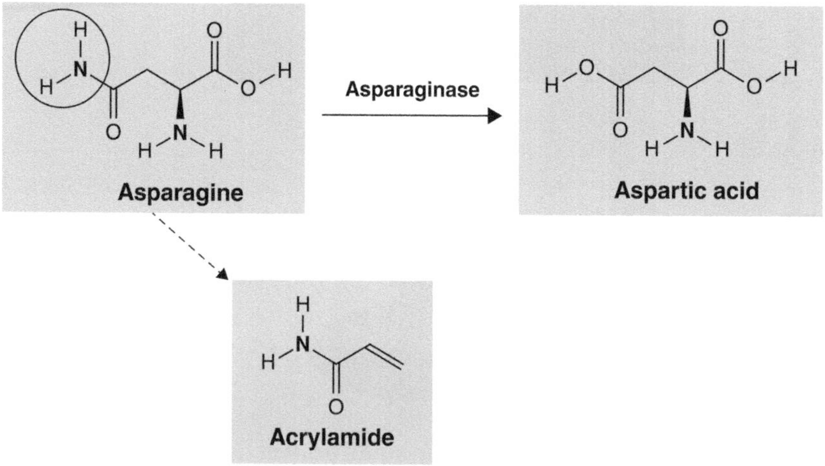

Asparagine

Asparaginase

Aspartic acid

Acrylamide

FIGURE 8.4 **Asparaginase therapy** converts asparagine, needed by rapidly growing cancer cells, to aspartic acid, thereby reducing their growth. Pretreating vegetables with the enzyme can also reduce (deaminate) acrylamide, a breakdown product of asparagine induced by treating foods at high temperature, to make safer potato chips and fried foods.

Another lytic enzyme, asparaginase, which converts the amino acids asparagine to aspartate (Figure 8.4), has been part of treatment for acute lymphocytic leukemia (ALL) for about 70 years. Cancer cells can import aspartic acid but have a reduced ability to (re)make asparagine from it intracellularly in the amounts needed to support their rapid rates of protein synthesis. Asparaginase treatment effectively starves the cancer cells (just as its production by pathogenic bacteria starves normal cells during infection). The therapy has its drawbacks: normal cells can make their own asparagine, but are also negatively affected by lower serum levels. In addition, patients can develop an immune reaction to the enzyme. Originally, the enzyme used was isolated from guinea pig sera, but this was rapidly replaced by secreted asparaginase of *E. coli*. More recently, asparaginase secreted by *Erwinia chrysanthemi* (now called *Dickeya dadantii*, a bacteria causing plant soft rot which can harm the human gut) has been approved for patients who react adversely to the *E. coli* enzyme.[19]

In addition, the toxins produced by *B. pertussis* and *Corynebacterium diphtheriae,* bacterial infections that robbed the breath of so many children in past centuries, are now used to prevent disease. The extracellular toxins, once inactivated, are an efficient and inexpensive vaccine, which induces a strong antibody response. Use of individual vaccines or the combination DTaP led to a spectacular decrease in diphtheria, tetanus and whooping cough in the 20th century. The current DTaP vaccine includes inactivated acellular Pertussis toxins, together with inactivated toxoids produced by *Clostridium tetani*. The neurotoxin of the latter organism, also called tetanospasmin, induces the characteristic "lockjaw" and rigidity that leads to death in tetanus. Very low doses of this toxin have been reported to improve muscle growth in a placebo-controlled study in dogs with spinal cord injury, but there was no improvement in gait.[20] Application of this therapy in humans vaccinated against tetanus toxoid might be difficult unless it was injected into an immune-privileged area of the body.

There are many other types of toxins produced by bacteria which have valuable activities. For example, asparagine is a precursor of acrylamide (Figure 8.4) by a multistep process.[21] Blanching potatoes and adding *Acinetobacter* asparaginase (or that produced by other bacteria) can lower acrylamide in potato chips by up to 90%.

While the *Clostridia* that cause human diseases have been highlighted here, it should be noted that other members of this very large species secrete a variety of enzymes that degrade organic matter. These can be exploited for environmental and remediation use and their nitroreductases can be used in bio-butanol manufacturing.

Toxins Are Co-Expressed with Proteins that Allow Them to Penetrate Cell Walls

Highly evolved secretion systems that look like missiles or arrows play a role in pathogenesis of many toxins.[22] These projectiles, often referred to as syringes, have probably evolved to penetrate other microbes as well as to safely store toxins. Bacteria have honed the skills to build such complexes since the dawn of life, perfecting the ability of these systems, and endemic phages, to lyse the tougher, layered membranes of other species which, in the internecine fight for survival, may compete for precious nutrients and must thus be destroyed.

A type 3 secretion system (T3SS) is shaped much like a rocket, with most of the "injectisome" devoted to getting the toxin to its destination into the medium or into an attached cell. A T6SS (Figure 8.5) forms a cocoon of protein around a toxin molecule to introduce it into the cytoplasm of another cell to which the bacteria attach. Once the proteins surrounding it separate, the toxin is free to wreak havoc within the target cell. In the example, the target cells are much larger than the bacteria, as would be the case for a pathogen such as *V. cholerae* or *Acinetobacter baumannii* inserting toxins into a eucaryotic cell or ameba.

Pathogens such as *Mycobacterium tuberculosis* can use these systems to enter mammalian cells,[23] possibly giving them a temporary harbor from their continuous fight to survive.

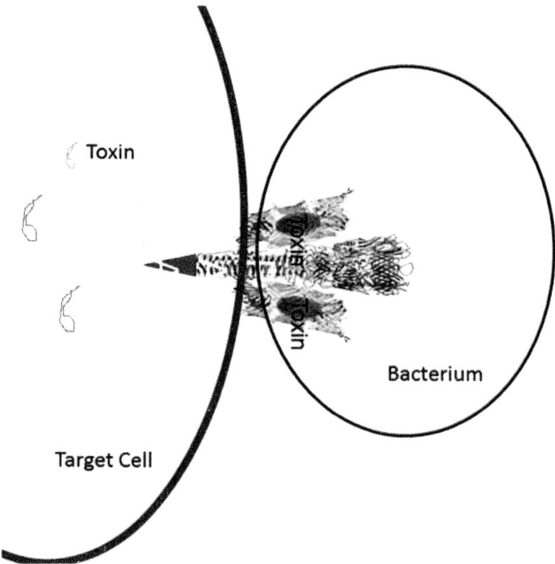

FIGURE 8.5 Like a guided missile, the toxin (small, enclosed ovals inside the complex, loops of protein after entering the cell) in a T6SS is encased in a structured complex of other proteins that inject it into the cytoplasm of a target cell. This injection system is usually much larger than its toxin payload.

Just as insulin's effects may be counteracted by glucagon, protein toxins in bacteria are often produced with an antitoxin. For example, the bacteria *Bacillus amyloliquefaciens* produces a nuclease called Barnase, which degrades RNA. However, it also produces a protein, Barstar, which binds to Barnase, inactivating it within the producer cell (Figure 8.6). For most of the thousands of toxins produced by bacteria, there is at least one "immunity protein" gene within the same bacterium encoding a small protein, able to bind to it and prevent it from acting.

As many toxins have very little activity unless they can get into the cytoplasm of the victim's cells, they are typically part of "AB" structures, where other components serve to facilitate entry into the target cell. Both anthrax toxins, for example, are dependent on the same pore-forming protein complex to get into cells and exert their activity. In a spectacular example of how scientific nomenclature can be misleading, *B. anthracis* produces a "protective antigen" (PA). This protein's name refers to the fact that it "protects" the toxins EF and lethal factor (LF) from degradation, while preparing a passage in human cells for them to enter and exert their toxic effects (Figure 8.3).

An inhibitor may even be part of the toxin itself. Most toxins are produced in a "pro" form, containing a domain that can be removed to

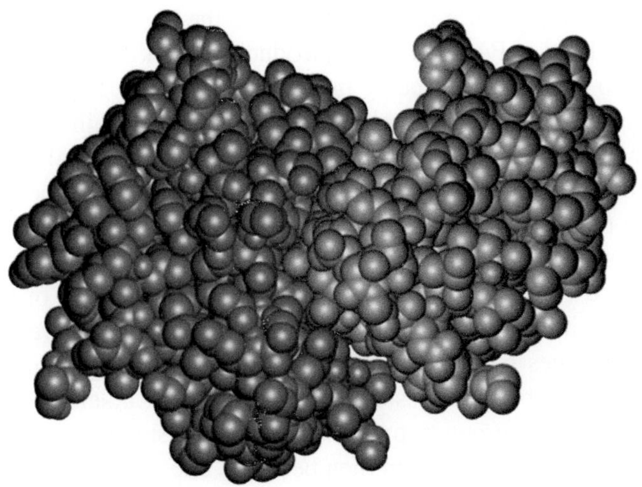

FIGURE 8.6 "Kissing proteins". The ribonuclease toxin Barnase (blue) is co-produced with its inhibitor, Barstar (pink). The two proteins form a very tight complex that prevents the toxin from killing the cell that produces it (structure from 1b2s.pdb).

generate an active toxin. Often, the enzymes that can cleave the prodomain are near the surface of the producing cell, and the cleavage occurs during the transport to the medium (in the case of secreted toxins). Alternatively, as with the anthrax toxin trio, the PA binding domain, PABD (see Figure 8.3) which is similar in both EF and LF, is cleaved during their transport into the cell, while they are within the channel made by PA. Thus, the individual toxins are synthesized with a prodomain, the PABD, to protect the producing cells or spores of anthrax, while also being incorporated into a multiprotein complex with the PA that allows them to enter foreign cells.

Bioinformatics and cloning has enabled the identification of similar toxins and their related toxin inhibitors in many different organisms. Toxins such as the nuclease Yoe B are typically controlled by labile antitoxins (YefM) (Figure 8.1). As toxin genes are induced in persister strains, identifying stable antitoxin homologues could have uses in developing mechanisms for controlling antibiotic resistant organisms.

Using toxin transporting molecules for protein and drug delivery and vaccines

The PA protein of anthrax toxins (Figure 8.3) forms a pore in the target cell's membrane to introduce the toxins into its cytoplasm. The PA can also help other proteins or large molecules to enter cells, if they are modified to contain the PABD region that interacts with it. Thus, PA, normally a part of the toxin complex of a pathogen, may be the basis of novel therapies. The potential use of PA combos in introducing cell-killing molecules specifically into cancer cells is being investigated.

Cholera toxin and its near relative, the heat-stable toxin of *E. coli*, are useful for studying signaling pathways.[24] They are additional examples of the AB family, where A is the enzymatic portion, while the B subunit (similar to the PA of anthrax toxins) allows cell interaction and access to the cytoplasm.

Inactivated extracts of these toxins, given in the absence of the bacteria, can alert the immune system to produce antibodies to prevent future invasion. Vaccines based on PA or the entry-permitting subunits of the other toxins have been suggested for Anthrax and Cholera. One obvious therapeutic use for immunity proteins would be to stop toxins in the blood of a sepsis patient. The problem then becomes one of identifying inhibitor molecules that bind closely to the toxins. The protein-encoding regions of natural inhibitors are generally near the toxin gene, typically encoded together on plasmid DNA. Alternatively, modern docking methods can also be used to select peptides capable of binding to a given surface of a toxin or its cell receptor. Selecting a protein capable of binding to a given surface is indeed a challenge, when starting only from the linear sequence of the protein.[25] However, recent advances in viewing protein binding pairs with high-resolution cryo-electron microscopy make the design of small inhibitory synthetic proteins[26] much easier.

Toxins Allow Bacteria to Evade the Immune System of Their Hosts

Release of bacterial toxins, LPS, nucleic acids and other products into the human cell cytoplasm activates the immune response. Some of this response, intended to protect the body, is counterproductive to both the host and the bacteria looking for a home. Rising body temperature and other effects of IL-1 can make the infected person feel a lot worse (see Chapter 6). For the bacteria, upregulation of cellular proteases (caspases and metalloproteases) and a cytokine storm raging in the blood can be deadly, so they benefit from having ways to protect themselves from the immune response, usually by hiding completely or shielding specific epitopes.

The most successful pathogens wrap themselves in polysaccharide sheathes, and sometimes lipids they steal from the host cells. In the most spectacular example, tuberculosis bacteria enter a cell and replicate there, somehow preventing all the defenses the cell has from being turned on to recognize their presence. Others produce proteins that bind to immune factors and stop them from signaling until induction of cytokines and autophagy mechanisms are already leading to the cell lysis needed to enhance bacterial spread.

Pertussis bacteria persist in whooping cough by preventing themselves from being recognized by neutrophils, which can form sheets around them. They express a protein on their cell surfaces, the "O" antigen, that allows them to enter respiratory epithelial cells where they will not stimulate phagosomes. In addition, bacterial adenyl cyclase toxins may interfere with generation of reactive oxygen species (ROS), a killing mechanism used by immune cells.

Immunosuppressive toxins

Bacteria produce many human-specific toxins that suppress the immune response, by blocking T-cell and neutrophil activation and preventing the release of inflammatory cytokines.[27] Even the powerful Stx has some immunosuppressive properties. The ability of these toxins to modulate the immune response and induce tolerance means engineered versions might be useful in treating autoimmune disorders and perhaps for controlling cytokine storm.

The first tissue transplants were enabled by a small protein produced by a fungal soil isolate that allowed it to evade the immune system. This protein, cyclosporin A, is tiny but requires a 40-step synthesis, by a coordinated series of enzymes, for the organism to produce it. While cyclosporin for therapeutic use is largely replaced by a synthetic small molecule, FK506 (Tacrolimus), the target of both is the mammalian protein calcineurin, a key factor in regulation of the immune response.

Toxins Can Slow the Growth of Their Producing Bacteria, Generating Antibiotic Resistant "Persisters"

As noted above, most lethal proteins are typically produced along with an intracellular "immunity protein" or an "anti-toxin" that can bind the toxin and neutralize it. When nutrients are plentiful and cells grow rapidly, these proteins neutralize the effect of the toxins and allow the attacking cell to resist the effects of its own toxins. When nutrient levels drop, the antitoxin is degraded and the toxin is free to wreak havoc even in the producing bacteria, stopping or slowing its growth. At any time, a certain number of cells in the culture may be in this toxin-induced, suspended state. Normally, these "dropouts" are a small number compared to the actively growing bacteria and would never be seen. But when an antibiotic enters the picture, most of the rapidly growing bacteria, those in which the antitoxin is doing its job, are killed. The only cells remaining are very slow-growing "persisters", held in this state by the released toxins or the presence of cAMP.

When more positive growth conditions return, or ATP levels recover, these cells gradually come to life and grow again unimpeded.

One of the molecules telling bacteria when they should stop growing is cAMP. cAMP levels generally rise as ATP falls with the energy level of the cell (Figure 8.2). Levels of cAMP tell bacteria when it is safe to grow actively, or when it is better to lay low and turn down their metabolism until the nutritional situation improves. The stalled cells enter a state much like hibernation in animals. These persister bacteria can essentially turn off their normal metabolism and uptake of nutrients from the environment completely. Antibiotics target actively growing cells but cannot affect the persisters. When the antibiotic is gone (and all those competitor bacteria along with them), the persisters slowly resume growth. The resulting re-infection may be even more severe as there is no longer the normal, protective flora to slow their growth.

However, simply giving more antibiotic to stop the regrowth may not help. For the persisters continue with their pattern of slower growth, meaning that the antibiotic will wash out before they are completely killed. In addition, slower growing resistant bacteria may block import of antibiotics into their cytoplasm, due to having thicker cell walls or mutations in the bacterial membrane "porin" proteins that prevent them from binding to and admitting antibiotics into the cell. They may also have enhanced export of the antibiotic, through upregulation of the cytoplasmic systems that bind and move drugs back out into their surroundings. Many bacteria have developed (and share with one another through plasmids) a variety of enzymes that degrade antibiotics, called for example β-lactamases or carbapenemases (which degrade penicillin and many other antibiotics that contain a β-lactam ring). One other complicating factor is the host immune system, which may produce serum proteins that bind to antibiotics or even phages, preventing further antibacterial activity.

THE BOTOX STORY: THE MOST POTENT OF ALL NEUROTOXINS HAS MANY MEDICAL USES

Bacterial toxins can have many uses in therapy. As presented in Chapter 1, one of the reasons we want to understand how toxins work is that many of them have been shown to be very useful in medicine, or other areas of the biotech industry. As discussed under Storming Cytokines, TNF, a human protein that induces lysis of tumor cells, is induced by lipopolysaccharide (LPS) from the cell walls of bacteria. Dr. William Coley used this fact to produce the first anticancer, protein-based therapy. Due to its intrinsic toxicity, TNF (discussed in Chapter 6) has been difficult to use directly in chemotherapy. TNF inhibitors in turn have proved to be useful treatments for psoriasis, and other diseases where overproduction of TNF or proteins associated with this powerful cytokine occurs.

While we can easily explain the origin of many bacterial toxins as mutant forms of perfectly useful intracellular enzymes, or acquisition by the transfer of DNA from another organism as being important for bacterial survival, the botulinum neurotoxin (BoNT) discussed for the rest of this chapter provides no clear advantage to its producer. Neurotoxins are produced by the bacteria within the intestine but travel to and incapacitate the neuromuscular control of the infected host. BoNT, now marketed under various brand names, including Botox, Dysport, Xeomin, Myobloc and Jeuveau, can bind presynaptically to cholinergic nerve terminals, blocking the secretion of acetyl choline and signals to the muscles (Figure 8.7). During botulinus infection, this can lead to rapid death of the host, which also does not seem to be of advantage to the bacteria. However, when dealing with the bacterially produced toxins in this chapter, it is good to remember that the damage caused by any toxin depends on how much is produced, and how it is administered.

In the early days of canning, many people died from eating food contaminated with the gram-positive, spore-forming anaerobic bacteria, *Clostridium botulinum*,[28] which produces the potent and lethal BoNT complex of

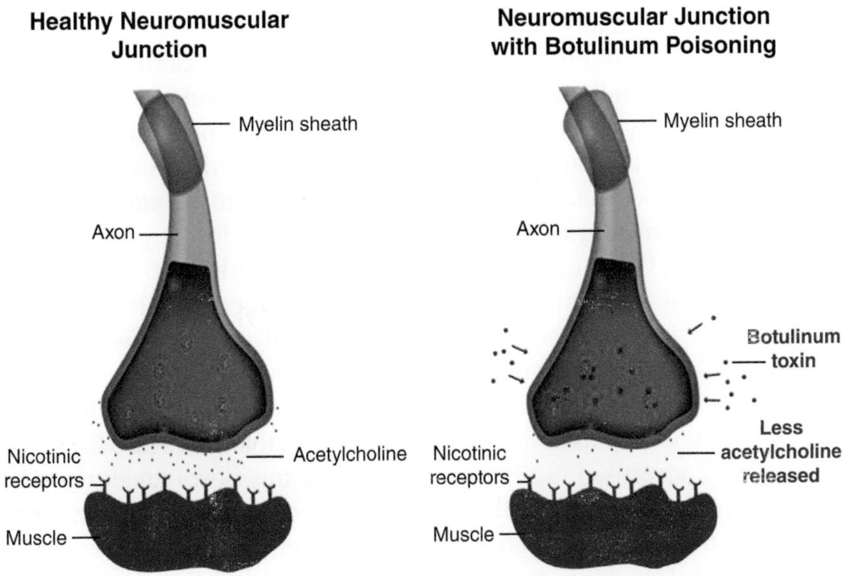

FIGURE 8.7 Mechanism of botulinus toxin poisoning. BoNT can move from the intestine to neurosynapses, stopping the signals that neurons send to muscles. This can lead to paralysis, even impairing the ability to swallow and breathe.

proteins. *C. botulinum* spores can survive heating and regenerate into neurotoxin-producing cells, which grow in improperly sterilized canned food. The major problem in canning is to preserve the quality of the food, while making sure that dangerous organisms are completely killed. Swollen cans indicate that the sterilization process has not been adequate, and that the food inside can be toxic. In the early part of the 20th century, canned food frequently spoiled on shelves and many cookbooks recommended that home canners heat their jars to just below boiling for at least an hour, trying to preserve nutrients in the food while still killing dangerous pathogens. However, Charles Olin Ball, after finishing studies in mathematics, delved into what it really took to kill off the spores of *C. botulinum* (anthrax bacteria also produce spores which can live for decades in soil). The resulting Ball curves revolutionized canning technology and the food industry in general. He showed that to completely kill spores, one needed very high temperature, which one could only achieve if the cans were placed under pressure. Pressure cooking allowed one to heat to 121°C, well above the boiling point of water at atmospheric pressure, for times as short as 20 minutes. This also preserved food quality better than the old methods of heating for hours at a time. His findings (as well as the introduction of pressure-stable jars and cans) led to pressure cookers in home kitchens and high pressure process lines in factories. Dried beans could go from the wash into a completely safe can in under an hour.

Nowadays, one really needs to make a series of catastrophic decisions to die from botulinum poisoning. First one needs to eat canned food that has been incompletely sterilized, a near impossibility with modern manufacturing controls, where any mistakes should clearly be seen by a swollen can. Further, the can would need to be a neutral food, such as green beans or meat (botulinum comes from the Latin word *botulus*, meaning sausage), not tomatoes or sauerkraut, as *Clostridia* do not grow well in acid conditions. One would need to eat the food directly from that swollen can, as BoNT is heat labile, meaning that cooking food completely should inactivate it.* However, botulinum infections still occur occasionally in those eating raw food contaminated with spores and infants fed raw honey.

While infections with *C. botulinum* are now rare, the much-feared BoNT is now widely used in medicine and cosmetology (Figure 8.8). The most widely advertised use is local injection of tiny doses under the skin (where it is not likely to use the neuropathway described in Figure 8.7 to inhibit central nervous system responses) to temporarily erase wrinkles. Similarly, local injection into the armpits reduces sweating and hence body

* John Steinbeck, in **Tortilla Flat**, describes how eating badly canned green beans kills Mrs. Morales chickens, but no one eating the cooked chickens has noticeable symptoms.

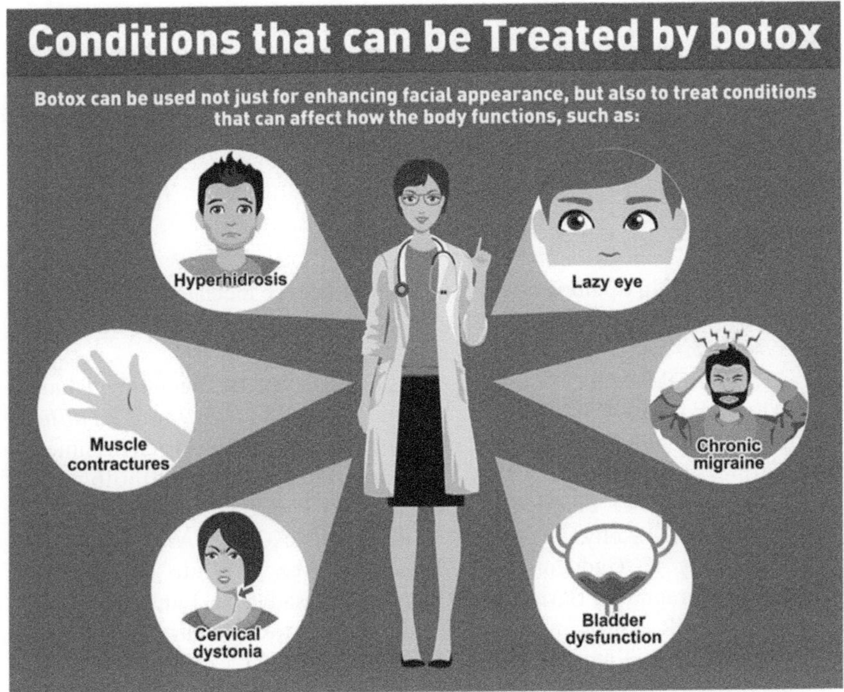

FIGURE 8.8 Low-dose BoNT has numerous uses in medicine and cosmetology.

odor. But the medical uses of BoNT injections are much wider. They are now used to alleviate migraine headache and chronic pain conditions,[29] upper body spasms, crossed eyes (strabismus), eyelid spasms and overactive bladder. In dentistry, injections may help ease jaw pain by relaxing the temporomandibular joint (that lies between the jaw and the skull).

The potential medical uses of dangerous toxins do not stop with BoNT or the other examples mentioned.[25] Engineered forms of the toxins of anthrax and their homologues have broad antitumor activities and suppress solid tumor growth in animal models.[30] Toxins that enhance cAMP production may also have anticancer activity. Cancer cells have very high metabolic rates, making them dependent on high levels of ATP. When cAMP levels rise in cells, they activate PKA, a protein kinase that is very important for activating other proteins by phosphorylating them. This can greatly inhibit the growth of cancer cells, if the toxin can be directed to them.

CONCLUSION

Our challenge is now not just to find antidotes to the many molecules we consider toxins, in the context of how they help pathogenic bacteria grow in the body, but also in how to use the myriad specific activities of these conditionally toxic proteins. Their uses will increase in the future, as we know more about the molecular and genetic basis of illnesses we now call "idiopathic".

REFERENCES

1. Poole K. Efflux pumps as antimicrobial resistance mechanisms. Ann Med. 2007;39(3):162–76. doi: 10.1080/07853890701195262. PubMed PMID: 17457715.
2. Jo J, Gonzales-Luna AJ, Lancaster CK, McPherson JK, Begum K, Jahangir Alam M, Garey KW. Multi-country surveillance of *Clostridioides difficile* demonstrates high prevalence of spores in non-healthcare environmental settings. Anaerobe. 2022;75:102543. Epub 20220225. doi: 10.1016/j.anaerobe.2022.102543. PubMed PMID: 35227896; PMCID: PMC9197859.
3. Alonso CD, Kelly CP, Garey KW, Gonzales-Luna AJ, Williams D, Daugherty K, Cuddemi C, Villafuerte-Galvez J, White NC, Chen X, Xu H, Sprague R, Barrett C, Miller M, Foussadier A, Lantz A, Banz A, Pollock NR. Ultrasensitive and quantitative toxin measurement correlates with baseline severity, severe outcomes, and recurrence among hospitalized patients with *Clostridioides difficile* infection. Clin Infect Dis. 2022;74(12):2142–9. doi: 10.1093/cid/ciab826. PubMed PMID: 34537841; PMCID: PMC9258941.
4. Gundolf T, Kalb R, Rossmanith P, Mester P. Bacterial resistance toward antimicrobial ionic liquids mediated by multidrug efflux pumps. Front Microbiol. 2022;13:883931. Epub 20220519. doi: 10.3389/fmicb.2022.883931. PubMed PMID: 35663893; PMCID: PMC9161554.
5. Terrinoni M, Nordqvist SL, Kallgard S, Holmgren J, Lebens M. A novel non-antibiotic, lgt-based selection system for stable maintenance of expression vectors in *Escherichia coli* and *Vibrio cholerae*. Appl Environ Microbiol. 2018;84(4). Epub 20180131. doi: 10.1128/AEM.02143-17. PubMed PMID: 29222103; PMCID: PMC5795084.
6. Hurlburt B, Lloyd SW, Grimm CC. Comparison of analytical techniques for detection of geosmin and 2-methylisoborneol in aqueous samples. J Chromatogr Sci. 2009;47(8):670–3. doi: 10.1093/chromsci/47.8.670. PubMed PMID: 19772743.
7. Zwe YH, Yadav M, Ten MMZ, Srinivasan M, Jobichen C, Sivaraman J, Li D. Bacterial antagonism of *Chromobacterium haemolyticum* and characterization of its putative type VI secretion system. Res Microbiol. 2022;173(3):103918. Epub 20211211. doi: 10.1016/j.resmic.2021.103918. PubMed PMID: 34906677.
8. Basi-Chipalu S, Sthapit P, Dhital S. A review on characterization, applications and structure-activity relationships of *Bacillus* species-produced bacteriocins. Drug Discov Ther. 2022;16(2):55–62. Epub 20220423. doi: 10.5582/ddt.2021.01087. PubMed PMID: 35466124.

9. LaRock DL, Russell R, Johnson AF, Wilde S, LaRock CN. Group A *Streptococcus* infection of the nasopharynx requires proinflammatory signaling through the interleukin-1 receptor. Infect Immun. 2020;88(10). Epub 20200918. doi: 10.1128/IAI.00356-20. PubMed PMID: 32719155; PMCID: PMC7504964.
10. Sha J, Wang SF, Suarez G, Sierra JC, Fadl AA, Erova TE, Foltz SM, Khajanchi BK, Silver A, Graf J, Schein CH, Chopra AK. Further characterization of a type III secretion system (T3SS) and of a new effector protein from a clinical isolate of *Aeromonas hydrophila* – part I. Microb Pathog. 2007;43(4):127–46. Epub 20070518. doi: 10.1016/j.micpath.2007.05.002. PubMed PMID: 17644303.
11. Zwack EE, Chen Z, Devlin JC, Li Z, Zheng X, Weinstock A, Lacey KA, Fisher EA, Fenyo D, Ruggles KV, Loke P, Torres VJ. *Staphylococcus aureus* induces a muted host response in human blood that blunts the recruitment of neutrophils. Proc Natl Acad Sci U S A. 2022;119(31):e2123017119. Epub 20220726. doi: 10.1073/pnas.2123017119. PubMed PMID: 35881802; PMCID: PMC9351360.
12. Bielaszewska M, Mellmann A, Zhang W, Köck R, Fruth A, Bauwens A, Peters G, Karch H. Characterisation of the *Escherichia coli* strain associated with an outbreak of haemolytic uraemic syndrome in Germany, 2011: a microbiological study. Lancet Infect Dis. 2011;11(9):671–6. doi: https://doi.org/10.1016/S1473-3099(11)70165-7.
13. Russell AB, Wexler AG, Harding BN, Whitney JC, Bohn AJ, Goo YA, Tran BQ, Barry NA, Zheng H, Peterson SB, Chou S, Gonen T, Goodlett DR, Goodman AL, Mougous JD. A type VI secretion-related pathway in Bacteroidetes mediates interbacterial antagonism. Cell Host Microbe. 2014;16(2):227–36. Epub 20140725. doi: 10.1016/j.chom.2014.07.007. PubMed PMID: 25070807; PMCID: PMC4136423.
14. Chen D, Ma L, Kanalas JJ, Gao J, Pawlik J, Jimenez ME, Walter MA, Peterson JW, Gilbertson SR, Schein CH. Structure-based redesign of an edema toxin inhibitor. Bioorg Med Chem. 2012;20(1):368–76. Epub 20111116. doi: 10.1016/j.bmc.2011.10.091. PubMed PMID: 22154558; PMCID: PMC3251925.
15. Chen D, Martin ZS, Soto C, Schein CH. Computational selection of inhibitors of Abeta aggregation and neuronal toxicity. Bioorg Med Chem. 2009;17(14):5189–97. Epub 20090527. doi: 10.1016/j.bmc.2009.05.047. PubMed PMID: 19540126; PMCID: PMC2743868.
16. Chen D, Misra M, Sower L, Peterson JW, Kellogg GE, Schein CH. Novel inhibitors of anthrax edema factor. Bioorg Med Chem. 2008;16(15):7225–33. Epub 20080628. doi: 10.1016/j.bmc.2008.06.036. PubMed PMID: 18620864; PMCID: PMC2678011.
17. Mueller AL, Brockmueller A, Fahimi N, Ghotbi T, Hashemi S, Sadri S, Khorshidi N, Kunnumakkara AB, Shakibaei M. Bacteria-mediated modulatory strategies for colorectal cancer treatment. Biomedicines. 2022;10(4). Epub 20220401. doi: 10.3390/biomedicines10040832. PubMed PMID: 35453581; PMCID: PMC9026499.
18. Yamamoto M, Yasunaga H, Kakinoki R, Tsubokawa N, Morita A, Tanaka K, Sakai A, Kurahashi T, Hirata H, CeCORD J Study Group. The CeCORD-J study on collagenase injection versus aponeurectomy for

Dupuytren's contracture compared by hand function and cost effectiveness. Sci Rep. 2022;12(1):9094. Epub 20220531. doi: 10.1038/s41598-022-12966-z. PubMed PMID: 35641603; PMCID: PMC9156707.

19. Salzer WL, Asselin BL, Plourde PV, Corn T, Hunger SP. Development of asparaginase *Erwinia chrysanthemi* for the treatment of acute lymphoblastic leukemia. Ann N Y Acad Sci. 2014;1329:81–92. Epub 20140805. doi: 10.1111/nyas.12496. PubMed PMID: 25098829.
20. Kutschenko A, Manig A, Monnich A, Bryl B, Alexander CS, Deutschland M, Hesse S, Liebetanz D. Intramuscular tetanus neurotoxin reverses muscle atrophy: a randomized controlled trial in dogs with spinal cord injury. J Cachexia Sarcopenia Muscle. 2022;13(1):443–53. Epub 20211027. doi: 10.1002/jcsm.12836. PubMed PMID: 34708585; PMCID: PMC8818617.
21. Xu F, Oruna-Concha MJ, Elmore JS. The use of asparaginase to reduce acrylamide levels in cooked food. Food Chem. 2016;210:163–71. Epub 20160422. doi: 10.1016/j.foodchem.2016.04.105. PubMed PMID: 27211635.
22. Costa TR, Felisberto-Rodrigues C, Meir A, Prevost MS, Redzej A, Trokter M, Waksman G. Secretion systems in Gram-negative bacteria: structural and mechanistic insights. Nat Rev Microbiol. 2015;13(6):343–59. doi: 10.1038/nrmicro3456. PubMed PMID: 25978706.
23. Russell AB, Peterson SB, Mougous JD. Type VI secretion system effectors: poisons with a purpose. Nat Rev Microbiol. 2014;12(2):137–48. Epub 20140102. doi: 10.1038/nrmicro3185. PubMed PMID: 24384601; PMCID: PMC4256078.
24. De Haan L, Hirst TR. Cholera toxin: a paradigm for multi-functional engagement of cellular mechanisms (Review). Mol Membr Biol. 2004;21(2):77–92. doi: 10.1080/09687680410001663267. PubMed PMID: 15204437.
25. Negi SS, Schein CH, Ladics GS, Mirsky H, Chang P, Rascle JB, Kough J, Sterck L, Papineni S, Jez JM, Pereira Mouries L, Braun W. Functional classification of protein toxins as a basis for bioinformatic screening. Sci Rep. 2017;7(1):13940. Epub 20171024. doi: 10.1038/s41598-017-13957-1. PubMed PMID: 29066768; PMCID: PMC5655178.
26. Schein CH, Levine CB, McLellan SLF, Negi SS, Braun W, Dreskin SC, Anaya ES, Schmidt J. Synthetic proteins for COVID-19 diagnostics. Peptides. 2021;143:170583. Epub 20210601. doi: 10.1016/j.peptides.2021.170583. PubMed PMID: 34087220; PMCID: PMC8168367.
27. Lopez Chiloeches M, Bergonzini A, Frisan T. Bacterial toxins are a never-ending source of surprises: from natural born killers to negotiators. Toxins (Basel). 2021;13(6). Epub 20210617. doi: 10.3390/toxins13060426. PubMed PMID: 34204481; PMCID: PMC8235270.
28. Jankovic J, Brin MF. Therapeutic uses of botulinum toxin. N Engl J Med. 1991;324(17):1186–94. doi: 10.1056/NEJM199104253241707. PubMed PMID: 2011163.
29. Pellett S, Yaksh TL, Ramachandran R. Current status and future directions of botulinum neurotoxins for targeting pain processing. Toxins (Basel). 2015;7(11):4519–63. doi: 10.3390/toxins7114519. PubMed PMID: WOS:000365647700011.

30. Peters DE, Hoover B, Cloud LG, Liu S, Molinolo AA, Leppla SH, Bugge TH. Comparative toxicity and efficacy of engineered anthrax lethal toxin variants with broad anti-tumor activities. Toxicol Appl Pharmacol. 2014;279(2):220–9. Epub 2014/06/28. doi: S0041-008X(14)00233-6 [pii] 10.1016/j.taap.2014.06.010. PubMed PMID: 24971906; PMCID: 4137396.
31. Chen D, Menche G, Power TD, Sower L, Peterson JW, Schein CH. Accounting for ligand-bound metal ions in docking small molecules on adenylyl cyclase toxins. Proteins. 2007;67(3):593–605. doi: 10.1002/prot.21249. PubMed PMID: 17311351.

Index

Note: Locators in *italics* represent figures and **bold** indicate tables in the text. Letter 'n' after locators refers to notes.

A

D

E

J

K

L

M

N

O

P

R

S

T

V

W

Y

Z

For Product Safety Concerns and Information please contact our EU representative GPSR@taylorandfrancis.com Taylor & Francis Verlag GmbH, Kaufingerstraße 24, 80331 München, Germany